ACS SYMPOSIUM SERIES **442**

Immunochemical Methods
for
Environmental Analysis

Jeanette M. Van Emon, EDITOR
U. S. Environmental Protection Agency

Ralph O. Mumma, EDITOR
Pennsylvania State University

Developed from a symposium sponsored
by the Division of Agrochemicals
at the 198th National Meeting
of the American Chemical Society,
Miami Beach, Florida,
September 10–15, 1989

American Chemical Society, Washington, DC 1990

Library of Congress Cataloging-in-Publication Data

Immunochemical methods for environmental analysis: developed from a symposium sponsored by the Division of Agrochemicals at the 198th National Meeting of the American Chemical Society, Miami Beach, Florida, September 10–15, 1989 / Jeanette Van Emon, Ralph O. Mumma, editors

    p.   cm.—(ACS Symposium Series, 0097–6156; 442)

Includes bibliographical references and index.

ISBN 0–8412–1875–7

1. Pollutants—Analysis—Congresses.   2. Immunoassay—Congresses.
3. Pesticides—Environmental aspects—Measurement—Congresses.
  I. Van Emon, Jeanette M., 1956– . II. Mumma, Ralph O., 1934– .
III. American Chemical Society. Division of Agrochemicals.
IV. American Chemical Society. Meeting (198th: 1989: Miami Beach, Fla.) V. Series

TD193.I48 1990
628.5—dc20   90–48309
                                CIP

The paper used in this publication meets the minimum requirements of American National Standard for Information Sciences—Permanence of Paper for Printed Library Materials, ANSI Z39.48–1984. ∞

ACS Symposium Series

M. Joan Comstock, *Series Editor*

1990 ACS Books Advisory Board

Foreword

THE ACS SYMPOSIUM SERIES was founded in 1974 to provide a medium for publishing symposia quickly in book form. The format of the Series parallels that of the continuing ADVANCES IN CHEMISTRY SERIES except that, in order to save time, the papers are not typeset, but are reproduced as they are submitted by the authors in camera-ready form. Papers are reviewed under the supervision of the editors with the assistance of the Advisory Board and are selected to maintain the integrity of the symposia. Both reviews and reports of research are acceptable, because symposia may embrace both types of presentation. However, verbatim reproductions of previously published papers are not accepted.

Contents

Preface ix

1. **Antibodies: Analytical Tools To Study Environmentally
   Important Compounds** 1
   Helen Van Vunakis

   IMMUNOASSAY EVALUATION GUIDELINES

2. **Immunoassays in Meat Inspection: Uses and Criteria** 15
   David B. Berkowitz

3. **Monoclonal Antibody Technology Program** 21
   Stephen Krogsrud and Kenneth T. Lang

4. **Development of Drug Residue Immunoassays: Technical
   Considerations** 27
   John J. O'Rangers

5. **Immunoassays in Food Safety Applications: Developments
   and Perspectives** 38
   Albert E. Pohland, Mary W. Trucksess,
   and Samuel W. Page

6. **Immunochemical Assays: Development and Use
   by the California Department of Food and Agriculture** 51
   Peter J. Stoddard

7. **Immunoassay Methods: EPA Evaluations** 58
   Jeanette M. Van Emon

   ACADEMIC ADVANCES IN IMMUNOASSAY TECHNOLOGY

8. **Polyclonal and Monoclonal Immunoassays for Picloram
   Detection** 66
   Raymond J. A. Deschamps and J. Christopher Hall

9. **Trinitrotoluene and Other Nitroaromatic Compounds: Immunoassay Methods** 79
   D. L. Eck, M. J. Kurth, and C. Macmillan

10. **Avermectins: Detection with Monoclonal Antibodies** 95
   Alexander E. Karu, Douglas J. Schmidt, Carolyn E. Clarkson, Jeffrey W. Jacobs, Todd A. Swanson, Marie L. Egger, Robert E. Carlson, and Jeanette M. Van Emon

11. **Immunochemical Technology in Environmental Analysis: Addressing Critical Problems** 112
   Bruce D. Hammock, Shirley J. Gee, Robert O. Harrison, Freia Jung, Marvin H. Goodrow, Qing Xiao Li, Anne D. Lucas, András Székács, and K. M. S. Sundaram

12. **An Enzyme-Linked Immunosorbent Assay for Residue Detection of Methoprene** 140
   J. V. Mei, C.-M. Yin, and L. A. Carpino

13. **Barriers to Adopting Immunoassays in the Pesticide Analytical Laboratory** 156
   James N. Seiber, Qing Xiao Li, and Jeanette M. Van Emon

IMMUNOASSAY ACTIVITIES IN INDUSTRY

14. **An Enzyme-Linked Immunosorbent Assay for Clomazone Herbicide** 170
   Ratna V. Dargar, John M. Tymonko, and Paul Van Der Werf

15. **Immunoassay Detection Methods for Alachlor: Application to Analysis of Environmental Water Samples** 180
   Paul C. C. Feng, Stephen J. Wratten, Eugene W. Logusch, Susan R. Horton, and C. Ray Sharp

16. **Competitive- and Inhibition-Type Immunoassay for Determination of Endosulfan** 193
   Bernhard Reck and Jürgen Frevert

17. **Monoclonal Antibody-Based Enzyme Immunoassay for Atrazine and Hydroxyatrazine** 199
   Jean-Marc Schlaeppi, Werner Föry, and Klaus Ramsteiner

18. **An Enzyme-Linked Immunosorbent Assay (ELISA) for Maduramicin in Poultry Feed** 211
     Rosie B. Wong

Author Index 221

Affiliation Index 221

Subject Index 222

Preface

IMMUNOCHEMICAL METHODS PROVIDE RAPID, sensitive, and cost effective analyses for a variety of environmental contaminants. Although many chemists recognize the potential of immunochemical methods for environmental analyses, they have not yet implemented this technology on a large scale. Concerns such as data quality, availability of specific antibodies, reagent stability, and methods evaluation must be addressed if immunochemical methods are to gain widespread acceptance and commercialization. The development of imunochemical methods is multidisciplinary, which may also impede understanding and acceptance. We hope that this book will fulfill the need for both a basic understanding of immunochemical methods and an update on important technological advances. The introductory chapter by Helen Van Vunakis of Brandeis University is an excellent overview of the use of antibodies as analytical reagents, providing thoughtful insight into exciting cutting edge research.

Other researchers, in academia and the chemical industry, are currently developing new techniques for analyzing environmental contaminants. These activities need to be shared and coordinated to avoid duplication of effort. The regulatory community must be well versed in the advantages and limitations of immunochemical methods. It must also provide guidance on objectives for achieving the data quality needed to implement immunochemical methods in regulatory and monitoring programs. Thus, there is a clear need for open communication among these diverse groups to advance immunochemical methods for environmental analyses. The goal of the symposium, and the book upon which it is based, is to introduce the topic to those unfamiliar with it, facilitate dialog, and stimulate interaction. It should also provide hope to its supporters that immunochemical methods can indeed be used for environmental monitoring. To that end, the editors of this volume have focused on new methods developed in the academic community, current advances and in-house uses within the chemical industry, the anticipated guidelines for regulation and monitoring of immunoassays, and the regulatory requirements of various government agencies, as well as their coordination in establishing and maintaining guidelines for accepting immunochemical methods.

Immunochemical technology is rapidly advancing in many areas: development of field-portable formats, specific antibody generation, detection systems, quality control and quality assurance measures, and new applications. It is not a panacea but should be used when deemed to provide the most appropriate analysis.

Immunochemical technology is rapidly advancing in many areas: development of field-portable formats, specific antibody generation, detection systems, quality control and quality assurance measures, and new applications. It is not a panacea but should be used when deemed to provide the most appropriate analysis.

The editors gratefully acknowledge the contributions of all the authors, including Helen Van Vunakis, Bruce Hammock, keynote speaker at the symposium, the reviewers for their valuable comments, and Linda Danner (U.S. EPA) for her assistance in the preparation of this volume. We also acknowledge E. I. du Pont de Nemours and Company, CIBA–GEIGY Agricultural Division, Hazelton Immunochemistry Division, Pierce Chemical Company, Westinghouse Bio-Analytic Systems, and Ohmicron Corporation for their generous financial assistance.

JEANETTE M. VAN EMON
U.S. Environmental Protection Agency
Las Vegas, NV 89193–3478

RALPH O. MUMMA
Pennsylvania State University
University Park, PA 16802

September 12, 1990

Chapter 1

Antibodies

Analytical Tools To Study Environmentally Important Compounds

Helen Van Vunakis

Department of Biochemistry, Brandeis University, Waltham, MA 02254

Sensitive, specific and rapid immunoassays are being used at an ever increasing rate in environmental analytical laboratories. Recent advances such as the production of antibody binding fragments in genetically engineered bacteria and the monitoring of hapten-antibody interactions in anhydrous organic solvents will be of practical use in the future development of immunoassays. By using the techniques of the molecular biologist, antibody fragments with the desired specificities and affinities may be obtained for labile and toxic compounds without the need to synthesize hapten-carrier molecules for immunization or to utilize the laborious process of monoclonal production.. Immunoassays for lipophilic water insoluble molecules may be carried out in anhydrous organic solvents thereby minimizing problems associated with their insolubility in aqueous media and generally increasing the application of antibodies as biosensors.

Competitive binding techniques in the form of radioimmunoassays (RIAs) were used to quantify insulin in plasma samples about thirty years ago.(1) Since that time various immunoassays for detecting hundreds of molecules of endogenous (e.g. hormones) and exogenous (e.g. drugs) origin have been described. The molecules being detected differ in size, in chemical and physical properties and in biological activity. General instructions for developing different types of immunoassays are available from several sources (2-6).

Despite the fact that immunoassays can be sensitive, specific and rapid, their application to detect compounds of environmental importance has been limited to relatively few laboratories (7-10). This symposium volume serves an important function since it focuses on the current status of the immunochemical methods being used for environmental analyses. Many of the accompanying papers consider the basic principles and essential steps in setting up immunoassays, i.e. the covalent linkage of small molecules to carrier proteins, immunization with these

0097–6156/90/0442–0001$06.00/0

conjugates to obtain antibodies, the development of assay procedure (e.g. RIA or enzyme-linked immunosorbent assay (ELISA)), determination of the specificity of the antisera and sensitivity of the assay. The validation of individual immunoassay procedures is emphasized.

This overview will consider some of the advantages and limitations of using immunochemical techniques to identify and quantify environmentally important agents. A few illustrative examples will be taken from methods used to quantify nicotine and its metabolites. We are interested in nicotine because it is pharmacologically active and the agent responsible for addiction to tobacco products. Once, also, it was used widely as a pesticide. Its catabolism in mammals is complex; some of the more common metabolites are shown in Figure 1 (*11,12*).

Since immunoassays utilize antibodies as analytical reagents, the challenge is to obtain antibodies specific for an individual compound that may be present in a milieu of structurally related and unrelated compounds. Antibodies can be produced by *in vivo* immunization (polyclonal antibodies), hybridoma technology (monoclonal antibodies) and by genetically engineered bacteria (antibody fragments with affinities for specific molecules). Some caveats pertaining to the production and use of such antibodies in immunoassay procedures will be presented.

Source of analyte

In general, analytes in environmental samples (e.g., water, soil, waste products, air) are more difficult to analyze than those that occur in physiological fluids (e.g., serum, plasma, urine, saliva). With most immunoassays utilized in the clinical and drug screening laboratories, aliquots of the physiological fluid are assayed directly (no pretreatment is required). The maximum volume of individual sample that can be analyzed is determined by the assay conditions. The assays are usually sensitive in the pg-ng range. Obviously, the direct monitoring of physiological fluids is possible only when the concentration of the analyte is within the detection limits of the immunoassay. Attempts to sensitize the assay by analyzing aliquots of physiological fluids larger than those indicated in the control experiments are to be avoided. Erroneous results may be obtained due to the non-specific interference with the antigen-antibody reaction. With analytes present at very low concentrations, it is sometimes necessary to concentrate (e.g. by liquid or solid extraction procedures) prior to assay in order to obtain reliable data. Alternatively, the assay can be made more sensitive to permit analyses of smaller quantities of a compound (*13*).

The potential usefulness of any analytical procedure should be assessed by carrying out control experiments in samples comparable to those collected for the field studies. Analytes in environmental samples are often determined at the ppm-ppb levels. While an immunoassay may appear to be specific, sensitive and reliable under the pristine conditions of the laboratory (e.g. when the test substance is in buffer), it may give false results with environmental samples that vary sufficiently (e.g. in pH, ionic strength, viscosity, solubility, humic content, etc.) to alter the antigen-antibody interaction or interfere with the monitoring systems. Samples collected at the point source where the analyte is most concentrated (e.g. at the

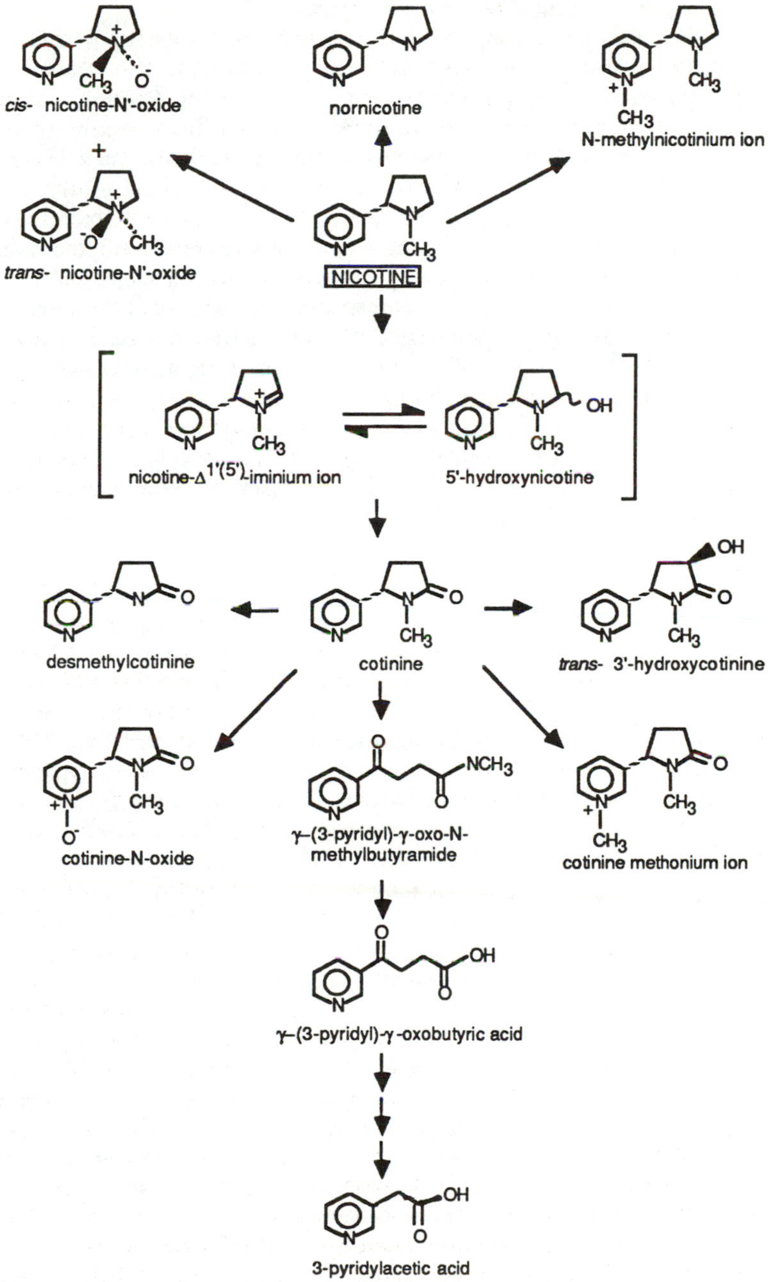

Figure 1. Pathways of metabolism of nicotine.

manufacturing site) may be assayed directly but after distribution (e.g. in ground water), cleanup and concentration are often required.

Agrochemicals that are highly lipophilic present additional problems. To favor solubility of such analytes, some immunoassays currently use combinations of aqueous and water miscible organic solvents in the medium. Recently, enzymes were shown to retain modified catalytic activities in anhydrous organic solvents (14). As a logical extension, the interaction of a small molecule (4 amino biphenyl) to its *immobilized* monoclonal antibody was tested in a variety of anhydrous organic solvents (15). The strength of the interaction was found to be related to the hydrophobicity of the solvent, i.e. the more hydrophobic the solvent, the weaker the antibody-ligand interaction. Obviously, antigen-antibody interactions differ and each system requires individual study to determine whether a specific compound can be analyzed under these conditions. For lipophilic water insoluble analytes, subsequent evaporation of the organic solvent used to extract and concentrate may not be required in an immunoassay being carried out in organic solvents. Thus difficulties encountered when an insoluble or "residue trapped" analyte is added back into an aqueous medium would be avoided. Also, the role of antibodies, as biosensors, may be significantly extended to measuring the antigen-antibody interaction in non-aqueous medium (15).

Antigens

Macromolecules such as proteins, nucleic acids, and polysaccharides usually can elicit an immune response when they are injected into the experimental animal directly or in the form of an electrostatic complex. Low molecular weight compounds (below the 5,000-10,000 range) ordinarily cannot elicit an immune response unless they are bound covalently to an antigenic macromolecule (16). A compound that is not immunogenic *per se* but can bind to its specific antibody is called a *hapten*. With haptens, the specificities of the hapten-antibody interaction can be predetermined to some extent by choosing the functional group that will be covalently bound to the carrier. The choice of utilizing a functional group already on the hapten or of introducing one for the specific purpose of forming the hapten-carrier conjugate depends upon the chemical structure of the hapten. If possible, it is better to retain the integrity of the original functional group(s) so that it can contribute to the hapten-antibody specificity. If, however, no functional group is present on the hapten (e.g. nicotine and cotinine), such a group can often be introduced at a desired position by synthetic procedures and subsequently used to covalently link the hapten to the macromolecule for immunization (17-19). Highly specific antisera were obtained for nicotine and cotinine using the hapten derivatives containing carboxyl groups (Figure 2). Immunoassays utilizing other hapten derivatives to prepare nicotine and cotinine antibodies have been reviewed (18, 19).

Because of the multiplicity of metabolites that can be formed during some biotransformation reactions, it is not always possible to obtain a specific antibody to a hapten. Antibodies that recognize a family of related molecules (i.e., parent compounds and closely related metabolites or environmental degradation products) are useful to determine total immunologically reactive material for screening purposes. Such was the case for the antisera prepared against dieldrin by the procedure outlined in (20). Aldrin and related organochlorine insecticides cross reacted to such an extent that the individual compounds could be quantified by immunoassay only after separation by an efficient chromatographic procedure (e.g., high pressure liquid chromatography). For some compounds, the synthesis of the hapten-carrier molecule necessary to obtain specific antibodies requires considerable skill on the part of the organic chemist.

Knowledge of the metabolism and stability of an agrochemical is essential. To detect exposure, it may be more useful to screen for a longer lived more abundant metabolite than for a rapidly catabolized or unstable parent compound. Often, specific antisera can be produced against individual metabolites. For example, cotinine (Figure 1) is the major product formed during the mammalian metabolism of nicotine (11,21). To determine exposure to tobacco products, samples are commonly assayed for this metabolite rather than the parent alkaloid (17,19,21). Nicotine has a relatively short half life in man (1-2 hr) and its concentration, even at peak times of smoking, rarely exceeds 50-60 ng/ml serum. Cotinine, on the other hand, has a relatively long half life (about one day) and remains relatively constant in habituated smokers (average is about 300 ng/ml serum) (22,23). Exposure to nicotine by the dermal route in non-smokers who harvest tobacco can also be monitored by analyzing their urine for cotinine (25). The physiological fluids of subjects who utilize nicotine gum in smoking cessation programs also contain cotinine. Urinary cotinine levels can also be used to detect exposure to other peoples' smoke (i.e. passive smoking) (26,27).

Antibodies can possess a high degree of specificity for a particular compound. Even with reagents as specific as the anti-nicotine and anti-cotinine sera, occasionally a non-related compound may cross-react for reasons that can not be ascertained by looking at its chemical structure. With the cotinine antisera, we have found one non-related compound that cross-reacts to a significant extent. Metyrapone, a cytochrome P_{450} inhibitor and a drug whose use is limited to testing the ability of the pituitary to respond to a decreased concentration of plasma cortisol, shows appreciable cross-reactivity with the anti-cotinine sera. Reduced metyrapone also cross reacts with the antisera. The fact that many thousands of physiological samples from non-smokers were negative for nicotine and cotinine when analyzed by this radioimmunoassay indicates that drugs and diets commonly used by these subjects do not react with these antibodies (19). However, the possibility that drugs and foodstuffs common to other cultures may contain cross-reactive material cannot be excluded. The need to check unusual findings by an independent technique is emphasized.

Antibodies

With a specific antibody, an analyte can often be quantified even in the presence of large amounts of extraneous materials that can obscure other detection systems (e.g., absorption or fluorescence spectra). Antibodies are glycoproteins found in the globulin fractions of serum and in tissue fluids. They are produced by vertebrates in response to the presence of an antigen, i.e., a substance that is recognized by the host to be foreign. Antibodies show a remarkable ability to bind selectively the antigen that stimulated their production. Their specificity may be regarded as comparable to that of an enzyme for substrate. They are relatively stable, soluble molecules with known chemical and physical properties. Procedures for their purification are available, but diluted antisera is often used (e.g., in fluid phase immunoassays). It has been estimated that an individual animal has the potential to produce antibodies specific for approximately 10^7-10^8 diverse immunodominant moieties. These molecules may possess binding constants for individual antigens on the order of 10^4-10^{12} M^{-1}. This ability of antibodies to discriminate between the homologous antigen and the myriad of other compounds of widely diverse structure that are found in experimental samples is of fundamental importance in their use as analytical tools (28,29).

Knowledge about the complex processes by which antibodies are produced *in vivo* is very extensive and, as yet, incomplete (28-31). Upon exposure to antigen, certain cells of the lymphatic system (the B-cells whose surface immunoglobulin can bind the antigen) are stimulated to proliferate and differentiate into cells that can secrete antibodies. The process is adaptive, it involves cooperation from other cells (e.g. T-cells and macrophages). The antibodies secreted by a single cell can be highly specific for a particular antigenic determinant. The serum collected from an immunized animal contains antibodies that are products of many stimulated clones. Different cell types in the various species can produce immunoglobulins that differ in size and carbohydrate content. Classes and subclasses of immunoglobulins exist that can be distinguished serologically, electrophoretically, by physical characteristics, chemical properties and biologic activities. Fragments of such molecules (e.g. Fab, Fc) can be obtained by chemical and/or proteolytic treatment of the intact antibody molecule.

Most polyclonal antibodies have been made in rabbits. Generally, rabbit antibodies specific for haptens and used in immunoassays are of the IgG type (Figure 3). Similar schematic representations of the Y-shaped IgG molecule are found today in even elementary texts devoted to the biological sciences. With large antigenic molecules (e.g. proteins) containing many antigenic determinants (epitopes), the antibody population represents products of several stimulated clones that are complementary to multiple epitopes on the antigen. Also, unlike many haptens, large proteins have defined conformations in the native (compared to the denatured) forms. Most antibodies would be specific for the native conformation if native proteins are used for immunization.

Most of the agrochemicals are relatively small molecules and the antibodies produced in animals may, by comparison, be fairly uniform with respect to complementarity. When antisera to haptens are diluted sufficiently so as to favor interaction with the most avid antibodies, the Scatchard plots often are indicative of fairly homogeneous populations of antibodies. Their affinity constants could reach as high as 10^{12} M^{-1}. The intermolecular forces involved in the binding of antigens to antibody include hydrophobic, Van der Waals, electrostatic and hydrogen binding (28-31).

The polyclonal antibodies obtained by immunizing experimental animals have been and will continue to be satisfactory reagents for many immunoassay methods. Choices can be made among adjuvants, routes of injection, dosage and immunization schedules, species of animal to be immunized, etc. (2-6, 28-31).

Different animals immunized with the same conjugate can produce antibodies that may differ in affinities, titer, and specificities. Such differences are apparent with antibodies studied by the more classical physical chemical procedures. For a particular immunoassay, each antiserum from an individual animal must be characterized separately to select those that have the proper affinities and specificities. A single animal (e.g., a rabbit) can furnish antisera for many thousand determinations depending upon the titer of the antisera, the affinities of the antibodies and the individual immunoassay procedure used.

In 1975, Kohler and Milstein described the production of monoclonal antibodies that originate from a single clone of a B-cell obtained from the spleen of an immunized animal (*32*). After successful fusion with a mouse tumor cell and appropriate screening, hybridomas can be isolated that produce antibody molecules with a single specificity. With multivalent antigens (i.e. proteins and other large molecules), cross-reactivity is markedly diminished because the specificity can be directed to a single unique epitope on these molecules. Monoclonals tend to have lower affinities. Polyclonal and monoclonal antibodies generally exhibit similar specificities for haptens of limited size. Monoclonals are homogeneous reagents and can be obtained in large quantities in ascites fluid. Proper maintenance of the hybridoma cell lines is required. Animals and tissue culture techniques are involved in their production. Initially at least, the methodology used to obtain monoclonals is costly and labor intensive.

Monoclonal antibodies are particularly useful for the production of stereospecific antibodies in cases where the immunizing conjugate contains a racemic derivative of a hapten. The antisera from animals immunized with such a conjugate would contain antibodies to both isomers. By hybridoma technology, antibodies that are stereospecific would be produced by individual clones. It is, of course, possible to develop assays that can quantify the individual isomers with the polyclonal antibodies provided the cross reactivity between the natural and unnatural isomer is negligible. For example, with rabbit antisera, the radioimmunoassays can detect the natural isomers of nicotine or cotinine because [^{3}H]-(*S*)-(–)-nicotine or [^{3}H-(*S*)-cotinine are used as ligands (*19*). The assay could be adapted to quantify the unnatural isomer of nicotine by using [^{3}H]-(R)-nicotine as ligand. Recently, several anti-nicotine and anti-cotinine hybridomas were selected by a screening procedure that utilized immunoprecipitation of the [^{3}H] labeled natural isomers of nicotine or cotinine to optimize selection of antibodies specific for these isomers. Stereospecific monoclonal antibodies for nicotine and cotinine in concentrations up to 7.5 mg/ml ascites and with binding affinities that generally exceeded 10^8 M^{-1} were obtained (*33*).

The theoretical concepts related to the idiotypic network and the nature of molecular mimicry are important areas of immunological research (*34-37*). Although the intricacies of the natural process are complex, several practical applications of anti-idiotypic antibodies to biological problems have been made, e.g. to detect and isolate certain receptors. The competitive inhibition by ligands of the idiotype-anti-idiotype interaction will also enable the environmental scientist to develop novel immunoassay procedures for compounds of interest. Two specific antibodies are required. The idiotype (Ab1), specific for the hapten, is obtained by immunizing animals with the hapten-conjugate.   Ab1 is then used as the immunogen to obtain antibodies (Ab2) specific to their combining sites (the anti-idiotypes). In each case, hybridoma technology and proper screening methods are used to select clones producing Ab1 (specific for the hapten) or Ab2 (specific for the hapten combining site on Ab1). The reaction between Ab1 and Ab2 can be inhibited by the hapten since it competes for the combining site on Ab1.

Recently, anti-idiotype antibodies were obtained using the monoclonal antibodies to (S)-(−) cotinine as antigen (38). An idiotype-anti-idiotype hapten immunoassay was developed which relies on the ability of cotinine to inhibit binding between a monoclonal anti-cotinine antibody (the idiotype) and a second monoclonal antibody (the anti-idiotype) specific for the antigen combining region on the idiotype. Because only monoclonal antibodies and antigen are required, this novel immunoassay obviates the need to prepare labeled hapten derivatives or macromolecular conjugates for solid phase assays. Immunoassays for proteins have also been described based on the inhibition of idiotype-anti-idiotype interaction (39).

Two laboratories have recently reported that antibody binding fragments could be produced in genetically engineered bacteria (40,41). These methods have the potential of replacing the hybridoma technology currently used for monoclonal antibody production. Advances in molecular biology, including utilization of the polymerase chain reaction (PCR) to amplify genes, were employed to clone the amplified genes in E. coli. Huse et al. (40) used separate libraries of genes that code for the light and heavy chains of the antibody and inserted them into a novel bacteriophage lambda vector system. A combinatorial library of Fab fragments of the mouse antibody repertoire were expressed in E. coli. The initial Fab expression library was constructed from mRNA obtained from a mouse that had been immunized with a p-nitrophenyl phosphoamidate-keyhole limpet hemocyanin conjugate. This hapten, a transition state analog, induced antibodies that can hydrolyze the analogous carboxamide substrate, i.e. catalytic antibodies. The ability to obtain clones that expressed Fab fragments specific for a hapten that possessed antigen affinities in the nanomolar range is relevant since many environmentally important compounds are small molecules. As stated by the authors (40), immunization of animals may no longer be necessary since the combinatorial repertoire obtained from a non-immunized animal may contain genes with the required specificities. It is possible to screen a million bacterial colonies per day, and choose those that are producing antibody binding fragments for a particular compound. Problems associated with the stability or toxicity of the antigen and the sometimes complicated synthesis required to prepare hapten-carrier conjugates for immunization would be minimized or eliminated.

By immunizing animals with specific hapten-carrier conjugates, we have been very successful in obtaining specific antibodies to nicotine, cotinine (17-19), γ-(3-pyridyl)-γ-oxo-N-methylbutyramide (18), N'-nitrosonornicotine (42), the nicotinamide nucleotide analogues of nicotine and cotinine (43) and, most recently, the intermediate formed during the nicotine to cotinine conversion (Figure 1) (44). We have failed, despite many attempts, to obtain antibodies to another metabolite of nicotine, i.e. nicotine-N'-oxide.   The hapten derivatives may have been

COOH
CHO
NH_2CH_3
$CH=NCH_3$
O
O
O
N
N
N
CH_3
CH_3OH, H^+
trans-(4'-Carboxycotinine)
CH_2OH
$COOCH_3$
$LiAlH_4$
CH_3
CH_3
$CH_2-O-C-(CH_2)_2-COOH$
CH_3

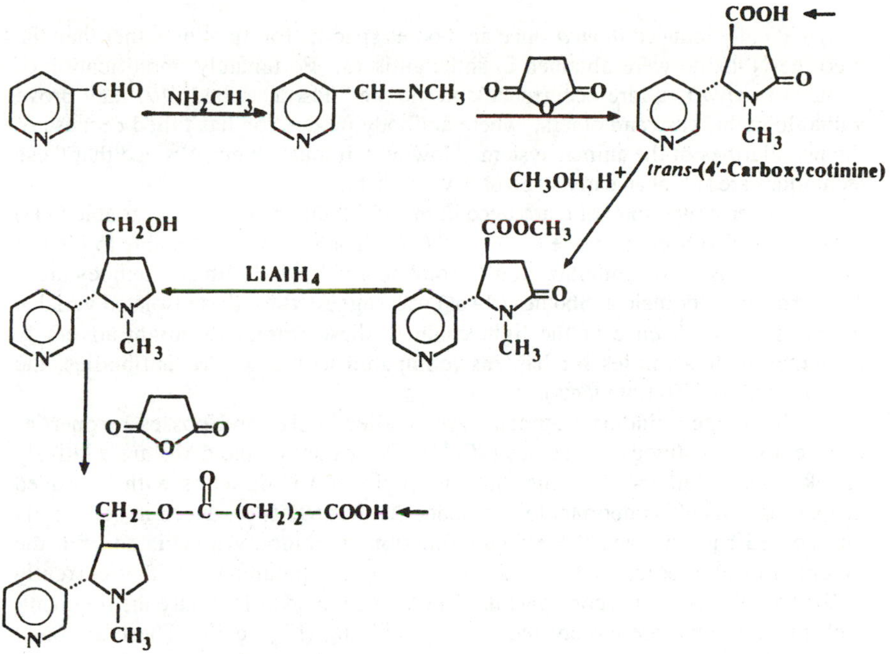

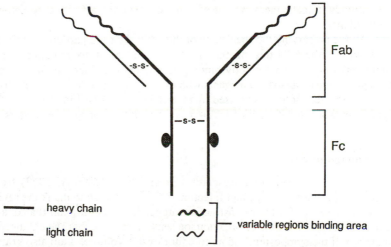

Figure 2. Preparation of hapten derivatives for conjugation to macromolecules.

-s-s-
-s-s-
-s-s-
Fab
Fc
heavy chain
light chain
variable regions binding area
Fab (fragment-antigen binding) – areas that determine binding to specific epitopes
Fc (fragment crystalline) – areas with other effector functions, e.g. contain carbohydrate moieties that determine binding to Staphylcoccal protein A
carbohydrate

Figure 3. Schematic structure of an IgG molecule.

enzymatically reduced *in vivo* since antibodies specific for nicotine rather than the nicotine-N'-oxide were obtained in some antisera. Fortunately, modification of haptens *in vivo* is a rare occurrence and the methods described (*40*) may prove valuable in this case and others, where antibody production has failed because of the peculiarities of the animal system. However, it must be emphasized that these techniques are still at an early stage of development.

A diverse repertoire of rearranged immunoglobulin heavy chain variable (VH) genes have also been expressed in *E. coli* (*41*). The products were able to bind to the proteins used for immunization (lysozyme and keyhole limpet hemocyanin). The term single domain antibodies (dAbs) was suggested for these isolated variable domains. The absence of the light chain in these fragments might adversely influence their affinities for haptens (compared to the *in vivo* antibodies, the monoclonals or Fab fragments).

The antigen-binding fragments are smaller in size and possess properties different from antibody molecules (*40-41*). For example, the dAbs are relatively "sticky" and attempts are being made to engineer VH domains with improved properties. It is also important to remember that fragments produced in genetically engineered bacteria lack the effector functions commonly associated with the Fc chain of antibodies. As a result, for example, immunoassays that currently utilize Protein A-enzyme conjugate in ELISA tests would fail because the fragments lack the carbohydrate moiety required for binding (Figure 3). These antibody binding fragments can not *a priori* be used interchangeably in immunoassays developed with polyclonal or monoclonal antibodies. However, the opportunity may exist to use site directed mutagenicity to engineer reactive functionalities into the antibody to make them more useful in biosensor and affinity chromatography applications.

While preserving the specificity of the antibody for a particular antigen, it is well within the abilities of the molecular biologist to enhance desirable characteristics of such antibody binding fragments (i.e. increase affinities, alter size) and to create analytical reagents with even more versatile properties. The rate at which this technology is utilized by other laboratories will in large part depend upon how, when and if the libraries are made available to other investigators.

Acknowledgments

H.V.V. is the recipient of a Research Career Award (5K6-AI-2372) from the National Institute of Allergy and Infectious Disease. I wish to thank the reviewers of this manuscript for their constructive suggestions. Additional thanks are due Jeanette Van Emon for the effort, common sense, and grace that she exhibited as an organizer of the symposium and as co-editor of this volume. I am also grateful to Bruce Hammock for consenting to give the keynote address in addition to the other scientific presentations that he had scheduled for the meeting. Publication no. 1709, Department of Biochemistry, Brandeis University, Waltham, MA 02254, U.S.A.

Literature Cited

1. Yalow, R. S.; Berson, S. A. *J. Clin. Invest.* **1960**, *39*, 1157.
2. Weir, D. M., Ed. *Handbook of Experimental Immunology, Fourth Edition*; Blackwell Scientific: Oxford, 1985; Vols. I and II.
3. Tijssen, P. *Practice and Theory of Enzyme Immunoassays. Laboratory Techniques in Biochemistry and Molecular Biology*; Burden, R. H.; van Knippenberg, P.H., Eds.; Elsevier: Amsterdam, 1985; Vol. 15.
4. Chan, D. W.; Perlstein, M. T., Eds. *Immunoassay. A Practical Guide* Academic: Orlando, 1987.
5. Chard, T. *An Introduction to Radioimmunoassay and Related Techniques. Laboratory Techniques in Biochemistry and Molecular Biology*; Burden, R. H.; van Knippenberg, P.H., Eds.; Elsevier: Amsterdam, 1987; Vol. 6, pt. 2.
6. Ngo, T. T.; Lenhoff, H. M., Eds. *Enzyme-Mediated Immunoassay*; Plenum: New York, 1985.
7. Hammock, B. D.; Gee, S. J.; Cheung, P. Y. K.; Miyamoto, T.; Goodrow, M. H.; Van Emon, J.; Seiber, J. N. *Pestic. Sci. Biotechnol., 6th Proc. Int. Constr. Pestic. Chem.*; Greenhalgh, R.; Roberts, T. R., Eds.; Blackwell: Oxford, 1987; p 309.
8. Van Emon, J. M.; Seiber, J. N.; Hammock, B. D.; Bruce, D. *Anal. Methods Pestic. Plant Growth Resul.* **1989**, *17*, 217.
9. Harrison, R. O.; Gee, S. J.; Hammock, B.D. In *Biotechnol. Crop. Prot. ACS Symposium Series*; American Chemical Society: Washington, DC, 1988; Vol. 379, p 316.
10. Mumma, R. O.; Brady, J. F. *Pestic. Sci. Biotechnol., 6th Proc. Int. Constr. Pestic. Chem.*; Greenhalgh, R.; Roberts, T. R., Eds.; Blackwell: Oxford, 1987; p 341.
11. Gorrod, J. W.; Jenner, P. *Essays Toxicol.* **1975**, *6*, 35.
12. Obach, R. S. Ph.D. Thesis, Brandeis University, Waltham, Mass., 1990.
13. Harris, C. C.; Yolken, R. H.; Krokan, H.; Hsu, I. C. *Pros. Natl. Acad. Sci. USA* **1979**, *76*, 5336.
14. Klibanov, A. M. *Chemtech* **1986**, *16*, 354.
15. Russell, A. J.; Trudel, L. J.; Skipper, P. L.; Groopman, J. D.; Tannenbaum, S. R.; Klibanov, A. M. *Biochem. Biophys. Res. Commun.* **1989**, *158*, 80.
16. Landsteiner, K. *The Specificity of Serologic Reactions*; *Second Edition*, Harvard University Press: Cambridge, Mass., 1943.
17. Langone, J. J.; Gjika, H. B.; Van Vunakis, H. *Biochemistry* **1973**, *12*, 5025.
18. Langone, J. J.; Van Vunakis, H. *Methods Enzymol.* **1982**, *84*, 628.
19. Van Vunakis, H.; Gjika, H. B.; Langone, J. J. In *Environmental Carcinogens Methods of Analysis and Exposure Measurement*; O'Neill, I. K.; Brunnemann, K. D.; Dodet, B.; Hoffmann, D., Eds.; IARC Scientific: Lyon, 1987; p 317.
20. Langone, J. J.; Van Vunakis, H. *Res. Commun. Chem. Path. Pharm.* **1975** *10*, 163.
21. Benowitz, N. L. *Ann. Rev. Med.* **1986** *37*, 21.
22. Benowitz, N. L. In *The Pharmacology of Nicotine*. ICSU Symposium Series. IRL Press: Oxford, 1987; Vol. 9, p 3.

23. Van Vunakis, H.; Tashkin, D. P.; Rigas, B.; Simmons, M.; Gjika, H. B.; Clark, V. A. *Arch. Environ. Health* **1989**, *44*, 53.
24. Zeidenberg, P., Jaffe, J. H.; Kanzler, M.; Levitt, M. D.; Langone, J. J.; Van Vunakis, H. *Comprehensive Psychiatry* **1977** *18*, 93.
25. Gehlbach, S. H.; Williams, W. A.; Perry, L. D.; Freeman, J. I.; Langone, J. J.; Peta, L. V.; Van Vunakis, H. *Lancet* **1975**, *478*.
26. National Research Council. *Environmental Tobacco Smoke. Measuring Exposures and Assessing Health Effects*; National Academy Press: Washington, DC, 1986.
27. U.S. Department of Health and Human Services. *The Health Consequences of Involuntary Smoking. A Report of the Surgeon General*; DHHS Publication No. (CDC) 87-8398; U.S. Department of Health and Human Services, Public Health Service, Centers for Disease Control: Washington, DC, 1986.
28. Kabat, E. A. In *Methods Enzymol*; Van Vunakis, H.; Langone, J. J., Eds.; Academic: New York, 1980, Vol. 70, p. 3.
29. Roitt, I.; Brostoff, J.; Male, D. *Immunology*; C. V, Mosby, St. Louis, 1985.
30. Nisonoff, A. *Introduction to Molecular Immunology*; Blackwell Scientific: Oxford, 1982.
31. Harlow, E.; Lane, D. *Antibodies. A Laboratory Manual*; Cold Spring Harbor Laboratory: Cold Spring Harbor, 1988.
32. Köhler, G.; Milstein, C. *Nature* **1975**, *256*, 495.
33. Bjercke, R. J.; Cook, G.; Rychlik, N.; Gjika, H. B.; Van Vunakis, H.; Langone, J. J. *J. Immunol. Methods* **1986**, *90*, 203.
34. Cleveland, W. L.; Erlanger, B. F. In *Methods in Enzymology*; Langone, J. J.; Van Vunakis, H., Ed.; Academic: New York, 1986; Vol. 121, p 95.
35. Linthicum, D. S.; Farid, N. R., Eds. *Anti-Idiotypes, Receptors and Molecular Mimicry*; Springer-Verlag: Berlin and New York, 1988.
36. Köhler, H.; Kaveri, S.; Kieber-Emmons, T.; Morrow, W. J. W.; Müller, S.; Raychaudhuri, S. In *Methods in Enzymology*; Langone, J. J., Ed.; Academic: New York, 1989; Vol. 178, p 3.
37. Kussie, P. H.; Albright, G.; Linthicum, D.S. In *Methods in Enzymology*; Langone, J. J., Ed.; Academic: New York, 1989; Vol. 178, p 49.
38. Langone, J. J.; Bjercke, R. J. *Anal. Biochem.* **1989**, *182*, 187.
39. Potocnjak, P.; Zavala, F.; Nussenzweig, R.; Nussenzweig, V. *Science* **1982**, *215*, 1637.
40. Huse, W. D.; Sastry, L.; Iverson, S. A.; Kang, A. S.; Alting-Mees, M.; Burton, D. R.; Benkovic, S. J.; Lerner, R. A. *Science* **1989**, *246*, 1275.
41. Ward, S. E.; Güssow, D.; Griffiths, A. D.; Jones, P. T.; Winter, G. *Nature* **1989**, *341*, 544.
42. Castonguay, A.; Van Vunakis, H. *Anal. Biochem.* **1979**, *95*, 387.
43. Shen, W. C.; Greene, K. M.; Van Vunakis, H. *Biochem. Pharmacol.* **1977**, *26*, 1841.
44. Obach, R. S.; Van Vunakis, H. *Drug Metab. Dispos.* **1990**. In press.

RECEIVED August 3, 1990

IMMUNOASSAY EVALUATION GUIDELINES

Chapter 2

Immunoassays in Meat Inspection
Uses and Criteria

David B. Berkowitz

Food Safety and Inspection Service, U.S. Department of Agriculture, Washington, DC 20250

The Food Safety and Inspection Service (FSIS) is developing criteria for the use of immunoassays in the meat and
poultry inspection program. Immunoassays and other new
analytical technologies present new options for the operation of the inspection system. Testing that once re
quired shipping samples to laboratories and expensive
laboratory procedures can now be done in processing
plants or on farms. Tests for the detection of drugs or
pesticide residues for animal diseases transmissible to
man, or for pathogenic bacteria are all of interest to
FSIS. FSIS is eager to use commercially available test
ing systems. Because the test results have impli-
cations for human health, FSIS is developing criteria
to be certain that the tests perform adequately and are
used in conjunction with quality control and quality
assurance programs.

The Food Safety and Inspection Service of the United States
Department of Agriculture (USDA) ensures that meat and poultry
products are safe, wholesome, and accurately labeled. In 1988, the
Agency inspected about 120 million head of livestock, 5.6 billion
birds, and 150 billion pounds of processed products.

Eight thousand inspectors stationed in plants across the
country inspect each carcass. To test for chemical residues
entering the food chain, inspectors collect statistically-directed
samples and send them from plants to one of three field laboratories
for analysis. Inspectors may also have samples analyzed on the
basis of questions about a product or for special surveillance
programs.

The meat inspection law was passed in 1906, largely in response
to the publication of The Jungle, Upton Sinclair's book describing
the unsanitary environment in which some food was produced at that
time. Since that time the general level of sanitation acceptable by
the meat and poultry industries and by society in general, has
greatly improved. The animals are more uniformly healthier.
Inspectors are far less often confronted with conditions that are

blatantly unsanitary. A greater proportion of effort is now directed toward detecting pathogens and chemical residues in the food supply. These cannot be detected by visual observation. Tests that can be performed at the inspection locations could greatly improve the effectiveness of inspection without increasing costs. For example, the estimated cost for the shipping and handling of each sample is about $75; this does not include the cost of the analysis itself. The overwhelming majority of samples tested are negative. On-site screening tests will enable the Agency to test many more samples, and to send a much higher percentage of positive samples to the field laboratories for confirmation. Additionally, rapid on-site tests have greatly expanded the spectrum of items that can be tested at inspection locations.

The increase in the number of tests performed is already large. For example, in 1980 the inspection program analyzed 200,000 samples. In 1988, FSIS analyzed 463,000 samples. In addition to increased numbers of samples tested, the spectrum of testing has also been enlarged. FSIS now monitors for E. coli 0157:H7 and Listeria which were not routinely tested 5 years ago. In the last year, the number of residue compounds tested increased from 112 to 120. Both the number and kinds of tests performed are increasing.

Advances in technology have made the increases in number and repertoire possible. The introduction of the radioimmunoassay (RIA) by Yalow and Berson in 1958 has been recognized by a Nobel Prize (1). The technology was made more assessable for inspection needs by the development of the enzyme-labeled immunosorbent assay (ELISA), by Engvall and Perlmann in 1971 (2). Many innovations since 1971 have resulted in portable, easy-to-use, disposable formats. The simplicity of the innovative formats for testing and the wide spectrum of applicable analytes accommodated by antibody diversity make immunoassays one of the most attractive technologies for the detection of disease, harmful chemicals and toxins, and economic fraud such as species substitution.

The FSIS Test Review System was developed to encourage the development of new tests by making the testing requirements known to test developers, and by expediting and standardizing the review process. All methods designed to measure the same analyte will be evaluated by the same standard, e.g., HPLC methods, microbiological inhibition tests, and immunoassays for the same analyte would all be evaluated from the perspective of limit of detection, accuracy, and precision.

The Test Review Process and Criteria

The review process contains 4 major milestones: a decision to review the test, a decision to accept the laboratory characterization of the test, a decision to accept the results of a collaborative study, and the final decision to approve the test. Each of these are discussed below.

The decision to review is necessary to ensure that FSIS resources are spent on reviewing only those tests that will be used for regulatory purposes by the Agency. The review process is resource intensive, and the FSIS could not afford to review every

test seeking USDA endorsement. In fact, FSIS is restricted by law
to spending money only to support the inspection program, and can
review only those tests applicable to the program. Conversely, test
developers must understand that approval does not guarantee purchase
or use. Inspection priorities change, so specific long-term needs
cannot be projected. For example, immunoassays specific for
sulfamethazine are now being considered. If FSIS finds that
producers are switching to other sulfonamides, the Agency would
continue to use the thin layer sulfa-on-site(SOS) test that detects
seven different sulfonamides. The decision to consider tests will
be made in the light of Agency needs at the time of the submission.
      The acceptance of the laboratory characterization of tests will
be based on the kinds of information usually included in a methods
development paper. FSIS suggests that supporting data be submitted
in a publication format or, as a reprint, if the information has
already been published. The information should include the purpose
of the test, the analyte and matrix (including the species), a
summary of the technique, its unique features, and performance
characteristics such as limit of detection, limit of reliable
measurement, and precision. The performance near regulatory
decision points is of particular importance. As would be expected
of a method development paper, the reagents, equipment, and the
details of the method must be thoroughly described.
      For yes/no tests, the evaluation of the false negative and
false positive rates at the decision point requires special
consideration. A standard procedure for evaluating yes/no tests has
not been established, but some of the considerations are compared
and contrasted with the evaluation of a quantitative test in the
following discussion. The discussion can be understood by referring
to Figure 1. The figure shows response vs. concentration curves for
a quantitative test (light dotted line) and for a yes/no test (dark
solid line). In the figure, the left ordinate is signal intensity,
such as Absorbance in a spectrophotometric assay. The right
ordinate corresponds with the dark line, and represents the
percentage of trials recorded as positive at each concentration. In
both the quantitative and yes/no tests, the test performance must be
described by the number of samples recorded as positive as the
concentration increases to and through the threshold concentration,
T. At the decision concentration, T, a standard quantitative assay
paper would describe the accuracy and precision at that point; i.e.,
for samples containing levels of exactly T, the mean and the
standard deviation would be determined.   If there is no bias, the
mean will be T. If the error at T is distributed normally, 50% of
the samples at that level will be false negatives. If the standard
deviation is known as a function of concentration above and below T,
one could predict the percentage of positives as the concentration
falls below T, and the percentage of positives as the concentration
increases above T. With a constant standard deviation, the false
negative rate above T would decrease faster with increasing
concentration as the sensitivity factor (slope of the signal vs.
concentration plot) increases. The mechanism for describing the
performance of yes/no tests is not as obvious.

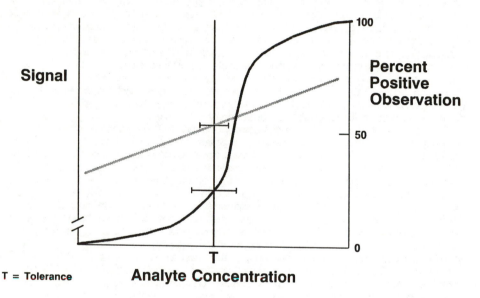

Figure 1.   Quantitative vs. Qualitative Assay Interpretation

For yes/no tests, the evaluation model is different. The analyte concentration at which 50% of the observers consider the test positive must be determined. The performance of the test can be described in terms of the percentage of tests recorded as positive by the trial participants at a series of concentrations below and above the threshold. The heavy line in Figure 1 is a hypothetical plot of the percentage of positive results at a series of concentration points. Unlike the quantitative test in which 50% of the samples at concentration T were positive, in the case illustrated by the heavy line, only about 20% of the samples of concentration T are positive. If the yes/no or positive/negative decision is made visually by a comparison with colored standards and the line does not cross the threshold concentration at or near the 50% positive point, one could adjust the standards to correct the concentration at which the 50% point occurs. Statistical methods for determining the concentration corresponding to the 50% point, the number of samples required at each concentration, and the behavior of the response curve below and above the 50% point have not been standardized. This is under consideration by an Association of Official Analytical Chemists' (AOAC) Task Force on Test Kits.

If the threshold is critical for health considerations, fewer false negatives would be permitted in either type of assay; it is likely in these cases that the threshold would be lowered. The false negative rate should rapidly decrease as the concentration increases above T. The number of false positives that can be tolerated will depend upon the regulatory consequences associated with positive results and the cost of the confirmatory tests.

Additional information required by FSIS should accompany the paper describing the laboratory characterization of the test. This is the type of information that would not usually be included in a scientific paper. This information is related to the planned use of the test that must be discussed with FSIS. The information includes cost estimates for installation and the cost per test at various levels of usage, the guarantee of a reagent supply, quality control measures for test operators, quality control measures for test manufacture, and a description of the training required for analysts.

The collaborative study should be conducted in the environment of the intended use of the test by people with the same level of training as the anticipated users. This kind of a collaborative study is designed to test performance, including the environment, training, and the quality control system. The test may fare better in a laboratory-based collaborative study, where defects could be dissected out and corrected. The test developer may choose to test the system in a mini-collaborative study based in laboratories, but before final consideration, FSIS would like to see the results of the collaborative study conducted on-site. These circumstances are likely to produce estimates of the test accuracy, precision, and bias that will be seen in the field.

The final step in the approval process is a consideration of all of the data supporting the test along with the details of the Agency's plan for use, Agency priorities, and the budget. If all of

the relevant issues seem compatible and there is still a need for
the technology, the test will be approved.

After the test is implemented, it will become part of a check
sample program that is operated by FSIS. Other quality assurance
measures may be taken by FSIS because tests are not expected to be
useful without adequate performance controls.

Discussion

All agencies have their own requirements for testing systems, but
whatever the special requirements, the evaluation of the performance
of a test should always be the same: statistical data supporting
specificity, sensitivity, accuracy, and precision are always
required. Discussions are underway with other agencies to provide
uniform requirements for this nonspecific part of the test approval
process.

Many tests designed for use by producers will not be reviewed
by FSIS because regulatory action will not be associated with the
results of testing. For example, producers may wish to test their
animals for residues before they are marketed. FSIS is encouraging
this kind of testing and is exploring mechanisms that might be
useful for advising producers of the reliability of commercially
available tests.

FSIS is interested in a wide range of analytical technologies,
including immunoassays, for use in laboratories as well as at
inspection locations. Gene probes, supercritical fluid extraction
and chromatography, and the use of computers for data acquisition
and transmission are now being used or are in trials. Robotics have
been incorporated into laboratory procedures, and many of the
routine analytical systems are constantly being updated.
Immunoassays have a great deal to offer, but other technologies are
also being pursued.

The details of the proposed FSIS Test Review System have been
published in the Federal Register (3).

Literature Cited

1. Berson, S.A.; Yalow, R.S. J. Clin. Invest. 1959, 38,1196–2016.
2. Engvall, E.; Perlmann, P. Immunochemistry 1971, 8,871–874.
3. Federal Register, 1989, 54,33920–33923.

RECEIVED May 22, 1990

Chapter 3

Monoclonal Antibody Technology Program

Stephen Krogsrud and Kenneth T. Lang

U.S. Army Toxic and Hazardous Materials Agency, Aberdeen Proving Ground, MD 21010–5401

The U.S. Army Toxic and Hazardous Materials Agency
(USATHAMA) has sponsored the development of methods
for analysis of tetryl, dieldrin, benzene, and
p-chlorophenylmethylsulfone using monoclonal antibodies.
While the work with tetryl has resulted in a test with
a detection limit of approximately 2 ppm, work is still
in progress to develop methods for the other three
analytes.

The U.S. Army Toxic and Hazardous Materials Agency, or USATHAMA, is
a Field Operating Agency of the U.S. Army Corps of Engineers that
offers a wide spectrum of environmental support to Army
installations nationwide. Services include conducting remedial
investigations and feasibility studies, as well as research into
new methods of waste minimization, remediation, and environmental
analysis. USATHAMA provides these services through contracts with
environmental engineering firms throughout the country. Research
projects, as well as routine analysis, are performed by contract
laboratories, since USATHAMA has no laboratory facilities of its
own.

Currently, USATHAMA is investigating or cleaning up
environmental problems at over 83 installations. These include
depots and equipment rebuild facilities, ammunition plants, and
installations listed in the congressionally mandated base closure
plan. In this work a wide variety of contaminates have been
encountered from sources such as plating sludges, degreasers, paint
and solvent wastes, and fuels/lubricants.

Standard U.S. Environmental Protection Agency methods are
generally used by USATHAMA to analyze environmental samples, but
these methods are not available for all compounds of interest.
Also, it is often desirable to have a method of analysis which has

a higher sample throughput than standard laboratory methods or which can be done quickly in the field. Analyses using monoclonal antibodies can fill these needs, as well as offering the possibility of high selectivity, low detection limits, and low cost per sample. As USATHAMA is not currently using immunoassay techniques in any of its projects, the decision was made to develop, as a trial, methods for four compounds which are of concern at various installations. These compounds are: tetryl (trinitrophenylmethlynitramine), benzene, dieldrin, and p-chlorophenylmethylsulfone. These compounds are shown in Figure 1. The goal is to produce a test for water samples which can be adapted to either lab or field use and can, with minimal sample preparation, measure target analytes in the low ppb range.

The tetryl work has been completed. Work continues at the present time on the development of analyses for the remaining compounds. The work is being performed by the organizations listed in the acknowledgements.

Approach

The problem of developing an immunoassay is basically one of isolating a monoclonal antibody with the required reactivity and specificity. These antibodies are produced by an animal's immune system in response to inoculation with an antigen. In the work reported here, the animals used were mice (Balb/c) or rabbits (New Zealand White). The antigen was, of course, different for the development of each antibody.

As the animal's immune system will not recognize a compound with a molecular weight as low as that of the compounds in Figure 1, it was necessary to prepare a hapten-protein conjugate, or immunogen. The haptens synthesized are shown in Figure 2, along with the proteins to which they were conjugated.

In general, it was desired to bind the hapten to the protein in such a way that the most "distinctive" portion of the hapten was exposed. Analogs of each target analyte were synthesized containing an acid group at the desired point of conjugation with the protein. This can be seen in Figure 2. It was realized that benzene, which lacks distinctive functionality, would provide the greatest challenge to finding a specific antibody.

Results

Although in most cases only one immunogen was prepared, for dieldrin three immunogens were prepared. A study was made of the three to determine which had the greatest hapten concentration. The study showed hapten:protein molar ratios of 8.5, 25, and 71 for the BSA, OVA and THY proteins, respectively. These ratios were obtained by infrared (IR) analysis as well as elemental chlorine analysis. Chlorine analysis was used as only the dieldrin molecule contains covalently bound chlorine. The conjugates with the highest hapten loadings, OVA and THY, were used for immunization in the dieldrin case.

Tetryl

Benzene

Dieldrin

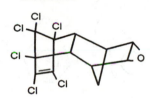

p - chlorophenylmethylsulfone

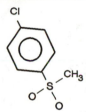

Figure 1.   Compounds Selected for Immunoassay Development.

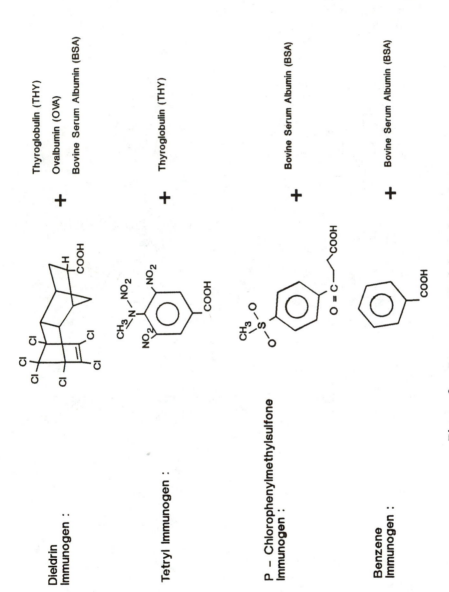

Figure 2. Immunogens Used in Each Project.

After a series of immunizations with the appropriate immunogen, lymphocytes were collected and fused to produce hybridomas. These were screened for antibodies with affinity for the desired analyte. A summary of the results is given in Table I.

Table I. Hybridoma Production

Analyte	Number of Fusions	Hybridomas Isolated With Positive Antibodies
Tetryl	2	16
Dieldrin	5	8
Benzene	8	0
p-Chlorophenyl-methylsulfone	9	1

As previously stated, cloning and antibody testing is complete for the tetryl analysis at this time, however work continues with the other three analytes. The lack of functionality undoubtedly has contributed to the difficulty of producing an antibody for benzene.

In the tetryl case, the most sensitive antibody was selected and used in an assay based on competitive inhibition. The detection limit in water of the method is approximately 2 ppm. The cross-reactivity of the antibody is given in Table II.

Table II. Tetryl Monoclonal Antibody Cross-Reactivity

Compound	Reactivity
Aniline	2 %
2,6-Dinitroaniline	< 1 %
2,4-Dinitroanaline	< 1 %
1,3-Dinitrobenzene	10 %
1,3,5-Trinitrobenzene	5 %
2,4-Dinitrotoluene	< 1 %
2,6-Dinitrotoluene	2 %
2,4,6-Trinitrotoluene	2 %
N-Methyl-2-Nitroaniline	< 1 %
N-Methyl-4-Nitroaniline	< 1 %

Conclusion

Because the detection limit achieved by the tetryl immunoassay is significantly higher than that obtainable by more conventional methods, such as HPLC, the assay will find its greatest use as a field screen rather than a laboratory method. We are hopeful that the work we are sponsoring for the other analytes will result in tests usable in the laboratory as well as the field.

Any decision to develop other new immunoassays will depend on the success of the three projects still underway. However, we will continue to monitor the development of immunoassays by both other government agencies and by industry to identify tests relevant to our mission. By adapting commercially developed immunoassays as well as funding methods development for specific compounds, we hope to keep our environmental analysis program as efficient and cost-effective as possible.

Acknowledgments

The work on developing the immunoassay for tetryl was done by Westinghouse Bioanalytic Systems Co., of Rockville, Maryland, while the dieldrin work is currently in progress at Battelle Memorial Institute in Columbus, Ohio. The work with p-chlorophenylmethylsulfone and with benzene is being done by Southern Research Institute, Birmingham, Alabama.

RECEIVED April 16, 1990

Chapter 4

Development of Drug Residue Immunoassays
Technical Considerations

John J. O'Rangers

Division of Chemistry, Center for Veterinary Medicine,
U.S. Food and Drug Administration, Rockville, MD 20857

The growth in routine applications of biochemical methods of analysis began in 1959 with the introduction of radioimmunoassay by Berson and Yalow in the United States and Roger Ekins in England. Since its introduction in 1959, radioimmunoassay (RIA) has evolved from an esoteric research technique to a widely used tool in laboratory medicine. Although radioimmunoassay will continue to be extensively used, the rate of growth of the RIA market appears to be slowing with the so-called "non-isotopic" or more accurately, "non-nuclidic" immunoassays coming into wide use. The commercial outlook for these non-nuclidic procedures is excellent.

The balance of this paper will deal with a set of technical considerations that are generally applicable to the development and evaluation of biochemical screening methods. These technical points are not specific evaluation criteria since the evaluation of a specific test depends on the intended use of the data developed by the test. The technical points that follow exemplify the test performance information that should be available for all screening tests. This information should be of great help to a potential test user in deciding whether a specific test will be suitable for a defined need.

In the veterinary diagnostic area, it is estimated that the market for antibody based veterinary diagnostic tests will rise from the current 1.5 million per year to over 60 million by mid-1990. It is particularly interesting that of the 105 new biotechnology companies that have been formed, 40% are in hybridoma/monoclonal antibody technology research and development, with the largest commitment being in the pharmaceutical area. It is expected that not only will new pharmaceutical products be forthcoming but new analytical methods based on these biochemical technologies will also be commercially available. Several test kits for animal drugs and residues such as chloramphenicol, sulfa drugs and beta-lactams are already in the marketplace and are intended to be used as rapid screening methods.

Rapid screening tests offer many potential advantages in disease diagnosis, drug development and residue monitoring. However, the reliability of the residue information provided by the rapid tests can be compromised if the test methods exhibit inconsistent performance. In many cases, residue screening methods will be used in field, factory or farm environments by persons not specifically trained in chemical analysis and associated quality assurance procedures. The results from these tests are frequently forwarded to the Food and Drug Administration (FDA) either for potential regulatory action or as an alert to an alleged new drug residue or chemical contamination problem in regulated commodities. It is essential that rapid screening test methods used in these milieus be suitably evaluated. This evaluation should encompass both method performance assessment and assurances that the methods can be manufactured to a consistent quality.

Regulatory Situation

Although screening tests can be based on any aspect of analytical technology, most of the screening tests being introduced to the marketplace are based on immunoassay or biological receptor technologies. The products are being offered in a "test kit" format similar in configuration to those medical devices marketed for the diagnosis of human disease or other conditions. This has resulted in the public perception that animal drug residue test procedures must be regulated by the FDA, in a manner similar to human diagnostic devices. This is exemplified by the inquiries and requests that the Center for Veterinary Medicine (CVM) has received for information on the performance and use of rapid test kits. However, there is a substantial difference in FDA's regulatory authority for human in vitro diagnostic devices and animal drug residue screening tests, and it is important that this difference be clearly understood.

Under current United States law, there is no FDA pre-market approval requirement for a commercial screening test intended for the detection of animal drug residues; however, animal drug screening test kits are subject to post-market regulation. This authority is specified in section 201(h)(2) of the Food, Drug and Cosmetic Act (FDCA). This authority allows the FDA to regulate residue test kits as animal devices using sections 502(a) and 502(f)(1) of the FDCA to ensure that such kits' labeling is truthful, accurate and not misleading and bears adequate directions for use of the kit.

The situation described above pertains to animal drug residue screening tests or kits that are marketed as animal medical devices. If the screening test was included as part of the analytical methods for animal residues submitted to FDA in a New Animal Drug Application (NADA), the method could be subjected to testing and performance evaluation. The legal authority to require suitable analytical methods as a condition of drug approval is contained in section 512(d)(1)(a)(c)(h) of the Food and drug Act and in the Code of Federal Regulations (CFR), specifically, 21 CFR

514.1(b)(7). These regulations give FDA the authority to regulate and evaluate analytical methods, including screening methods, as part of the animal drug approval process. The evaluation of NADA analytical methods by FDA is to ensure the availability of suitable analytical methods that can reliably measure animal drug residues and indicate when the edible tissue of a treated animal does not contain hazardous drug residues.

The intended use of analytical methods in general and screening tests in particular, not only determine FDA's regulatory capabilities, but also determine the extent and complexity of the performance evaluation of the method. For example, in practice rapid screening tests can be used to assay animal tissues or fluids for specific drugs which are **known** to be present in the animal or to test animal tissue or fluids for the presence of drugs which **may** be present in the animal.

If we accept that the purpose of a screening test is to exclude the presence of a specific substance or substances, the validation requirements for the two types of tests described above differ chiefly in the demonstration of the degree to which the test method responds specifically to the test substances and in the need for confirmatory results.

If a rapid screening test is used to assay for the drug or residue that is known to be present in the animal, they key performance element is the specificity of the assay response for the test analyte. A typical use of a rapid screening test in this mode would be in a pharmacokinetic study to determine the half-life of a drug in the blood of a treated animal. The assay used should be able to track the concentration of drug with acceptable specificity and provide a kinetic assessment that reflects the effects of only the animal drug under test.

Independent confirmation of screening tests used to assay for substances **known** to be present in the animal is usually not needed as a routine part of the analytical test protocol, if the assay is specific for the test drug. If the test does not have sufficient specificity or is subject to interferences due to animal variability or treatment regimen, confirmation of screening test results can be performed, but is an inefficient procedure. It would be preferable to adapt the primary screening assay to achieve the required specificity or resistance to interferences. While the confirmation of assay results should be part of the basic validation work performed during the development of the screening test, confirmation need not be a routine part of the application of the screening test in a case where the drug being monitored is known.

The use of screening tests to determine if a specific drug or drugs are present in the tissue or fluids of an animal requires not only adequate test specificity, but may also require methods to confirm the screening tests results. The extent and complexity of the confirmatory testing needed will depend on the intended use of the analytical results. Testing that is performed for law enforcement purposes would require rigorous confirmatory methods such as gas chromatography–mass spectrometry, whereas testing for monitoring purposes could use less rigorous methods for confirmation, if confirmation is deemed necessary. The major point is that screening tests used in this mode require the availability of independent confirmatory methods. Every screening assay does not necessarily have to be confirmed, but the capability to confirm should exist. The need for confirmatory methods for screening tests used in this mode is actually implied in the definition of screening tests given above, i.e., a screening test can exclude the presence of the test compound. This means that where there is no **prior** knowledge that the test compound is actually present in the test sample, the results of the screening test only mean that the test substance **may** be present in the sample.

The interpretation of the term "maybe present" in terms of the statistical confidence in the screening test result depends on the performance characteristics of the assay and the verification of reliable performance by a body of actual test application data. Nonetheless, the confirmation that the test substance actually **is** present in the sample requires independent determination. For method development purposes, screening tests should be considered a part of an analytical system for a given substance, where the complete system consists of a separate confirmatory technique to verify the initial screening result.

The incorporation of screening or rapid tests into the NADA process for the purpose of regulatory application will routinely require that a rigorous confirmatory method be part of the analytical system. The requirement for a confirmatory method is part of FDA's Center for Veterinary Medicine policy for all analytical methods submitted for NADA regulatory purposes.

The technical points that follow exemplify the test performance information that should be available for all screening tests. The technical items are not intended as evaluation criteria since the evaluation of the performance of any specific test depends on the public health or regulatory significance of the test results and the ultimate use of the results. The technical points are intended to suggest to test developers the type of information that should be available for their products.

Technical Consideration in Screening Test Development

The Intended Use of the Method. The originator of the method should provide a description of the types of samples or matrices to which the method can be applied. Data should be available to demonstrate the application claimed for the method is supported by actual validation studies. The correct and effective use of a specific analytical method requires that the user understand the unique capabilities of the analytical method. This information will assist the user in determining if a specific test will be suitable for a particular application.

Ideally, the originator of the screening test will demonstrate the unique capabilities of the method by suitable experimental data. For example, if an immunoassay is developed to measure residues of animal drug "A" in a specific matrix and it is claimed that the immunoassay provides an estimate of the total residues present in the target tissue, it should be experimentally verified that the antibody fully cross reacts with the parent drug as well as all metabolites.

The characterization of the antibody, e.g., determination of affinity constants for parent drug and metabolites as well as the assay format used are important in the experimental data. However, the pivotal experiment in this example would be the feeding of test animals quantities of radiolabeled parent drug followed by the extraction and chromatographic separation of parent and metabolites from the target tissue and demonstrating that all radiolabeled chromatographic peaks fully cross react with the antibody used in the screening test procedure. This experiment would conclude with a comparison of the estimates of total residue determined chromatographically and by directly testing the unseparated extract.

Description of the Scientific Principles of the Method and the Critical Reagent or Instruments Used in the Method. The design of validation tests will depend on this information. For example, if uniquely derivatized solid phase high (HPLC) support was used in the method, this would be classified as a modification to existing technology. As such, the developer of the method should demonstrate that the technique of preparing the HPLC media is well understood, is reproducible and will yield batch-to-batch uniformity.

This point is especially important in evaluating analytical method that depend on biological reagents, such as enzymes, antibodies, receptors, etc. Although modifications to technology may not be involved, critical reagents are used that in themselves are subject to variability. It should be demonstrated that these reagents can be consistently produced and are widely available. It serves no purpose to collaboratively study a method which uses unique reagents of limited availability.

The information that can be deduced from the analytical response should be discussed. For example, a gas chromatography–mass spectrometry (GC–MS) method can give both quantitative and qualitative (identity) information about the test analyte. Note that the fragmentation pattern of a molecule in mass spectrometry is ultimately dependent on the chemical structure of the molecule. The structural identity of a test analyte can be deduced from this fragmentation pattern.

A method based on ligand binding, such as immunoassay, actually measures the response of an analytical system to the effects of the test analyte. In the case of immunoassay, the system response typically is the displacement of another molecule from the antibody binding site. The effectiveness of the displacement mechanism can be generally traced to some common structural domain shared by the competing molecules, but the displacement process may not give complete certainty of the structure of the test analyte. Where the competing molecules must share 100% of structural features for displacement from the antibody or receptor binding site, then very high specificity would be obtained. If the completing molecules share less that 100% functional identity (as is generally the case), then relative specificity occurs. This effect is also called "non-specificity," although this is a less accurate term for the analytical performance.

The point to keep in mind is that biological tests are, in general, refinements of a bioassay. Bioassays measure **function** directly and **structural identity** by inference. For example, antibiotics can be assayed by microbial inhibition assays. In these assays, the observed effect is the inhibition of microbial cell growth due to the effect of the antibiotic. In the case of immunoassays or other ligand assays, the observed effect is the competition or displacement of molecules from a binding site. Neither system **directly** determines the structure of the test analyte. While the performance of a test analyte in a well characterized bioassay or ligand assay can provide valuable presumptive information about the structural identity of the test analyte, these tests do not give definitive structural information. This is an important point in determining the best application of screening tests for forensic, public health or regulatory application.

Method of Synthesis and Characterization of the Critical Reagents. This item comprises the following:

- Specifications for evaluation of reagents
- Evidence of consistent manufacture of test systems
- Specifications of quality assurance tests

Quality assurance specifications should be defined by the developer of the test for all reagents, especially biological reagents. As outlined above, it should be shown that biological reagents can be consistently produced on a batch-to-batch basis. Performance specifications or specification range should be established for critical biological reagents. These performance specifications for specific reagents can be derived from the overall method performance specifications that are to be established. For example, in an immunoassay, the specificity, sensitivity and freedom from matrix interferences of the final test can be used to establish procedures and criteria for titer determination, specificity assessment, pH and ionic strength optima, etc., for the antibody, before the final test is assembled.

Certain types of equipment of materials that are critical to the performance of the assay should have quality assurance procedures established. For example, plasticware or glassware can absorb proteins or smaller molecules such as drugs. These absorptive effects can be useful, i.e., in developing a solid phase immunoassay wherein antibody is absorbed on the walls of the plastic or glass reaction vessel, or they can be a problem, i.e., by absorbing the analyte thereby making it unavailable for assay. It is important that these materials be monitored for acceptability in the assay, and recommendations provided to the user on how to assess the performance of glass or plastic material that are used in the analytical procedure.

Chromatographic procedures are often used in the extraction and preparation of an analyte(s) before assay. Procedures should be established for the chromatographic materials used to assure they will perform adequately.

Test System Logistics. This item deals with the manufacturing aspects of the test method or kit. Data should be available demonstrating that the methods of manufacture are well understood and in a state of control. It is important to know that the test characteristics can be maintained from batch-to-batch of critical reagent.

The development of specifications of each critical reagent or step in the manufacturing process and the determination degree of conformance with specifications is a practical way of establishing manufacturing control. The specifications should contain the acceptance or rejection criteria for the item tested and provide a reference for the technique for assessing the specification parameter.

Stability Data on the Critical Reagents. Test kit reagents and other components are typically lyophilized when feasible. Properly dried biochemical reagents are usually quite stable for extended periods, typically six months to one year or longer. However, when reagents are prepared for use, the lifetime of the reagents can be drastically reduced. This stability information should be available to preclude test malfunction due to degraded reagents. The user of the test should determine that the test materials or reagents will be stable under the actual conditions of use.

Stability Data on the Analyte(s). This data is especially important in drugs or chemicals occurring in biological matrices. Studies should examine the effects of matrix processing, e.g., cell disruption or tissue homogenization on the stability of the test analyte. This data will largely determine the methods for storing and shipping tissue samples containing the drug residue.

In addition to the effect of the physical state of the matrix on the stability of the analyte, the stability of the analyte after it is extracted from the matrix should be determined. This examination can be done at various steps in the analyte purification process if desired. Typical independent variables in stability studies are pH, temperature, light exposure, analyte concentration, storage vehicle and freeze/thaw cycles.

Determination of Assay Sensitivity with Replication and Statistics. Sensitivity can be defined as the ability of a test to discriminate between adjacent levels or concentrations of test analyte. There are other definitions of sensitivity, but the one specified is sufficiently general to serve several needs in residue analysis. For example, the definition recognizes that test sensitivity can vary with the point on the standard curve. If one of the points used is "zero," then the sensitivity estimate can be either the level of smallest quantitation or the level of detectability of the method. The intended use of the method will dictate the statistical requirements for the determination of the level of detectability or quantitation. Also, the definition allows a semi-quantitative reference level to be set for the test. This reference level could be the regulatory tolerance for a drug or marker residue.

Two sets of data should be developed to determine sensitivity. The first would be the determination of the analytical sensitivity of the method. This phase of the work is usually done in the laboratory using calibration standards and tissue or feed matrices that have been fortified or "spiked" with the test analyte.

The second phase is the determination of the number or percentage of true positive results achieved with the test in a population of animals that have been dosed with the compound of interest. This is an essential phase in the development of residue methods and the rigorous assessment of the true positive rate requires confirmation by a separate accepted assay method(s). In addition, part of this study may need to be performed under field conditions, particularly if the test is intended to be used in a non-laboratory environment.

The second phase of this sensitivity study requires the baseline variability in samples from residue free animals be known. This data will establish the negative control and will also indicate if components from a residue free matrix will interfere in the assay.

As a general rule, the level of detection of the assay will be no less than that of the negative control value plus three standard deviations. The actual level of detection should be verified experimentally.

The Variability Associated with Each Standard Point on the Analytical Curve. The reliability of immunoassay standard curves is not uniform across the entire dynamic range of the curve. The least analytical variability is usually observed in the central regions of the curve in the vicinity of 50% ligand displacement, with variability increasing at the extremes of the curve.

Evaluation of the validity of immunoassay measurements requires that the performance at various regions of the immunoassay curve be known. Ideally, this type of evaluation would involve measurement of sufficient replicate samples at each standard point on the curve. This would provide an assessment of the variability at each point. As a practical matter, measurement at the level of detection, 20% displacement, mid-range and the 80% displacement level of the standard curve would provide a reasonable assessment of performance.

For quantitative purposes, the region of the standard curve bracketed by 20–80% displacement should be used. This region tends to be the most linear and will provide the best accuracy and reproducibility.

Where an immunoassay is claimed to measure total residues, i.e., parent compound plus metabolites, standard curves generated using the parent analyte and each metabolite should be parallel. This is evidence that the same antibody is reacting with the analytes.

If an immunoassay is designed to measure one compound with high specificity, cross-reactants that give curves that parallel the curve for the analyte are not desirable. Parallel reactivity indicates that improvement in the assay cannot be achieved by purification of the antibody. Parallel cross reactivity in this case could make the antibody unusable for a specific assay.

Specificity of an immunoassay is usually measured by determining the extent that compounds that are structurally similar to the test analyte react in the assay. The determination of the assay reactivity of an array of potential cross-reactant is routinely performed in immunoassay development. A panel of suspect cross-reactant should be selected on the basis of structural similarities to the test analyte and on the expected occurrence along with the test analyte in the sample. Thus, an immunoassay specific for 17β-estradiol should be tested for reactivity with estrone, estriol, 17α-estraiol and testosterone. Other compound testing might also be indicated.

Data should be presented showing the number or percentage of true negative results obtained by testing samples from animals that have not been exposed to the drug or chemical. As in the determination of tests sensitivity, this evaluation may need to be performed in a field environment. However, unlike sensitivity, the determination of the negative rate in the estimation of specificity does not require a separate confirmatory analysis.

Test Samples. Samples should be selected to reflect the types of conditions that are likely to be encountered in everyday use of the test. There are three basic sets of samples:

> Set 1. Background or blank samples should come from animals that are known not to have been exposed to the test analyte. These samples will establish negative control.

> Set 2. Samples that are known to contain a definite quantity of the test analyte. Typically, these samples will be generated by "fortifying" or "spiking" a suitable matrix with the test analyte.

> Set 3. Samples that contain residues of the test analyte from the exposure of the animal or test matrix to the test analyte.

Ideally, these samples should be maintained in a test panel that can be used to evaluate different lots of a test kit. If a test panel is maintained however, the stability of the analyte under conditions of storage must be known.

The data on the true positive and true negative rates in conjunction with the drug use in the target population can be used to calculate the predictive value of the test. The predictive value is the percentage of true positives in a positive test population.

Interferences. Interference and specificity are closely associated but distinct concepts. Specificity is the extent to which the biological reagent, i.e., antibody, exclusively reacts with the test analyte. Specificity is very difficult to assess in absolute terms since specificity can only be measured by testing suspect cross reactant molecules in the analytical system. When the number of compounds that are potential cross reactants are finite and known, a good estimate of specificity can be determined. In order to make the number of cross reactants finite and known to a reasonable degree, the conditions and applications of the methods must be defined. Thus, specificity has a parochial meaning and does not have general meaning. The determination of specificity is even more difficult to determine with certainty in the forensic or residue area. In this case, the compound universe is not finite. Nonetheless, by carefully choosing the panel of compounds to be used in testing the specificity of a biochemical reagent, a useful estimate of its reactivity can be achieved.

Interferences, on the other hand, identify specific compounds or conditions that adversely affect the optimum performance of the method. Sources of interferences can be quite broad and diffuse. It is essential that the milieu in which the method is to be used be well understood so as to select potential interfering substances or conditions with some practicality and meaning. For example, if a rapid test is to be used on a farm environment, then the chemicals that typically are found in this environment should be tested for interference in the test.

The sample extraction and cleanup methodology can also be a source of interference, especially from solvent residuals or from adverse effects of processing on the analyte stability. In the case of analytes that are contained in complex matrices, e.g., feeds or tissues, a sufficient number of matrix blanks must be run to determine what the population blank really is and whether there are any interferences to be expected from this source.

Validation Studies. At a minimum, all validation studies should provide information on the following areas:

- Optimization of performance
- Identification of critical steps
- Recognition and control of interferences
- Assessment of method performance using authentic samples under authentic conditions
- Confirmatory analysis

There are several approaches that can be used for confirmation:

1. Use a definitive reference method to confirm the proposed method, e.g., confirm with mass spectrometry.

2. Confirm with several alternate tests. All test results should be consistent with each other and with the proposed method results. Where a definitive confirmation is not available or feasible, other methods can be used. In this case, more than one confirmatory test may be required to provide the certainty of identification. The confirmatory tests selected should be based on different principles from the pri-

mary test. For example, if an immunoassay is to be confirmed, multiple chromatographies under different conditions may be suitable.

3. Define an existing accepted method as a standard or reference and compare the proposed test with it.

Conclusion

There is no doubt that the speed of analysis and relative operational simplicity of these new screening tests offer many advantages to both FDA, drug manufacturers, clinicians, animal growers and ultimately to the consumer. In addition, third world nations could also benefit from the availability of drug and environmental screening tests for use in a first line screen of contaminants of public health concern.

While the scientific principles of biochemical analytical technologies are well understood and valid in principle, most of the practical experience in the use of these tests is in human laboratory medicine with most of the performance criteria meeting the analytical needs of this clinical sector. If immunoassay methods are to be usefully applied to the analysis of animal drug tissue residues, it will not be sufficient to extrapolate theoretical principles or data developed in human laboratory medicine to assess the suitability of immunoassay for regulatory or forensic analysis. It will be necessary to experimentally test immunoassay performance under conditions that are expected to be encountered in a monitoring or regulatory setting. These conditions can vary from "rough and ready" on-site field use to well-controlled laboratory environments.

The consistent manufacture and performance of current screening tests is a very important issue, however, much of the current information on the performance of screening tests is based on anecdotal information or personal experience of users. We need to have a well designed, objective evaluation of the state-of-the-art in the screening test area.

Any effort to develop performance criteria for screening test for animal drug residues must have the twin goals of providing for good method performance and of not unnecessarily restricting the development and maturation of an analytical technology that has potential benefits for human food safety and public health.

The recommendations in this paper are based on principles that are used in both drug residue analysis and in immunoassay method development. The issues discussed in this paper are intended to serve as a listing of points to consider for a developer or user of an animal drug screening test. No specific method performance standards are presented or method evaluation technique are presented. Specific criteria depend on the intended use of the analytical information. For example, method performance criteria for law enforcement are more stringent than criteria required for the establishment of trends in animal drug use. While the specific performance standard will vary depending on the intended use of the analytical information, each of the points outlined in this paper should have evidence suitable to demonstrate the validity of the screening test application.

RECEIVED August 7, 1990

Chapter 5

Immunoassays in Food Safety Applications
Developments and Perspectives

Albert E. Pohland, Mary W. Trucksess, and Samuel W. Page

Division of Contaminants Chemistry, Center for Food Safety and Applied Nutrition, U.S. Food and Drug Administration, Washington, DC 20204

Actual and perceived food safety concerns have necessitated an increase in the monitoring of foods for such natural contaminants as aflatoxins and for residues of pesticides. Immunoassays can provide rapid, simple, and relatively inexpensive methods for the detection of analytes with specificity and with sensitivities directed at the levels of concern. Particularly for aflatoxins, they are rapidly assuming a significant role in the monitoring of foods. However, the misuse of these techniques can potentially compromise any food safety improvement that may result from increased surveillance. Experiences of the Division of Contaminants Chemistry in the development, validation, and applications of immunoassays for natural toxins is discussed.

In the past few years there has been a keen interest in biotechnology-related research, and in the encouragement of the practical application of the results of such research, i.e., technology transfer. One area, of course, which has received a great deal of interest and attention involves the use of immunoassay techniques in the analysis of foods and feeds for a wide variety of contaminants for control and regulatory purposes. In this paper the evaluation of commercially available immunoassay kits in the analysis of foods and feeds for the mycotoxins and aflatoxins will be discussed. Aflatoxins B_1, B_2, G_1 and G_2 are highly toxic secondary fungal metabolites. Although <u>A. flavus</u> produces aflatoxin M_1, M_2 naturally, aflatoxin M_1, M_2 are more commonly found in excreta and milk of animals that have ingested aflatoxin contaminated feed. They are coumarin derivatives containing a fused dehydrofuran moiety (Figure 1). Aflatoxin B_1 is a demonstrated hepatocarcinogen in several animal species.

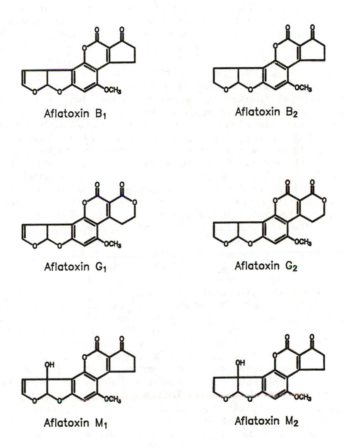

Aflatoxin B₁   Aflatoxin B₂

Aflatoxin G₁   Aflatoxin G₂

Aflatoxin M₁   Aflatoxin M₂

Figure 1. Chemical structure of the aflatoxins

It should be pointed out that at the present time none of the kits discussed are being used by the U.S. Food and Drug Administration (FDA) for regulatory purposes; however, we believe the potential for such use is great, especially in the sense of "screening" analyses. The conventional TLC, HPLC or mini column methods for detection of aflatoxins are time consuming and are not suitable for screening purposes. The Association of Official Analytical Chemists (AOAC) refers to a "screening method" as "a method to rapidly and reliably analyze a large number of samples at the designated level of interest in order to eliminate negative samples." (1) The U.S. Department of Agriculture/Federal Grain Inspection Service (FGIS) on the other hand, has approved the use of eight such kits in its control program (2); interestingly the USDA/Agricultural Marketing Service (AMS) has not agreed to automatically use the kits approved by FGIS (3).

In examining the kits one is immediately faced with the following concern: What criteria should be used in the evaluation of such kits? This is not a trivial question, and the number of opinions expressed on the subject indicates considerable diversity of thought. Because this question is central to decisions on eventual use of the kits, several organizations have addressed the issue by establishing committees assigned the task of developing evaluation guidelines. For example, the AOAC established an "Ad Hoc Task Force on Test Kits."(4) The International Union of Pure and Applied Chemistry (IUPAC) Commission on Food Chemistry has established a Working Group on Immunochemical Methods whose first project is to develop "Draft Guidelines on Criteria for the Evaluation, Validation and Quality Control on radioimmunoassay (RIA)-based Analytical Methods," to be followed by similar guidelines for enzyme-linked immunosorbent assay (ELISA) methods. The Environmental Protection Agency (EPA) has developed "Guidelines for EPA Evaluation Studies on Immunoassays" which are quite extensive (5). In the USDA similar guidelines have been developed and described in a document entitled "Design Criteria and Test Performance Specifications for Aflatoxin Screening Test Kits" (6). An announcement was made in the Federal Register describing the specific procedures the USDA intends to follow in approval of the use of test kits in its control programs(7). Various other groups are actively engaged in developing similar criteria.

In reviewing the efforts at establishment of criteria for evaluation of such kits, the following factors seem to be of concern:

USEFULNESS
- -Cost: reagents, equipment/facilities, speed of analysis
- -Use: laboratory/field, trained/untrained analyst (Referred to in the EPA guidelines as data quality objectives (DQOs).
- -Stability: shelf life
- -Quality assurance/Quality control: requirements, protocols, cost, availability of standards

CONFIDENCE FACTORS
- -Bias, precision, accuracy, specificity
- -Limit of detection/determination
- -Repeatability/reproducibility (within/between lab coefficient of variation)
- -False positive/negative results

Experimental Studies

There are currently at least six commercially available aflatoxin immunoassay kits produced in the United States, most of which have been extensively studied (Table I).

In addition, at least six foreign companies are marketing similar kits. In most cases these kits are dependent on generation of antibodies from the same hapten, i.e. aflatoxin B_1. Consequently, there is usually considerable cross reactivity toward other aflatoxin derivatives (B_2, G_1, G_2). It is important for proper evaluation of results obtained using the kits that this cross reactivity be known.

Our initial studies using the kits were disappointing, invariably resulting in modifications of the kits themselves and changes in protocols for use of the kits. Following this initial phase of work, we decided one kit was ready for collaborative study. This first collaborative study involved use of the Neogen Agri-Screen kit. The antibodies have specific ability to bind aflatoxin B_1 and very low cross reactivity to aflatoxins B_2, G_1 and G_2. This kit contains:

Antibody-coated microtiter wells
Aflatoxin standard solution
Dilution buffer (Tris)
Aflatoxin-enzyme conjugate (horseradish peroxidase)
Substrate A {2,2' azino-di[3-ethylbenz-thiazoline sulfonic acid]}
Substrate B (hydrogen peroxide)
Stopping solution (2N H_2SO_4)

In this collaborative study fourteen laboratories analyzed six different commodities containing aflatoxin B_1 at two concentration levels as blind duplicates. Two standards (15 and 50 ng B_1/g) were provided, and collaborators were asked to report their results as <15 or >15 ppb. Laboratories with microtiter well readers were asked to determine aflatoxin concentrations both visually and spectrophotometrically. In this kit aflatoxin B_1-antibodies are coated onto plastic microtiter wells. The aflatoxin-containing sample is extracted with $MeOH$-H_2O (55+45). The extract is defatted with hexane, and the MeOH extract mixed with the aflatoxin-enzyme conjugate and added to the well of the antibody-coated microtiter plate. The aflatoxin in the extract and the aflatoxin-enzyme complex compete for the antibody binding sites. The enzyme substrate(ABTS) and H_2O_2 solution are then added, the reaction leading to a colored product in the presence of enzyme. The intensity of color is determined visually or spectrophotometrically at 580nm.

The results of the collaborative study are tabulated in Table II (8).

Overall correlation (81%) between the ELISA instrumental results and those obtained using the AOAC recommended methods was good. On this basis the AOAC adopted the ELISA method official first action as a screening method for determining the presence or absence of aflatoxin B_1 at a concentration of ≥ 15 ng/g in cottonseed products and mixed feeds, a surprising conclusion in view of the relatively large numbers of both false positive and false negative results encountered. On the other hand, when an ELISA reader was used for quantitation, the RSD_R was <51% for cottonseed products and mixed feeds at levels >15 ppb; for all other commodities the RSD_R was

Table I. Commercially Available Immunoassay Test Kits
         for Aflatoxins

Manufacturer	Kit	Type	Analytes
Biomed Labordiagnostik (FRB)	Agromed	MTP[1]	B_1, M_1
Cambridge Life (UK)		MTP	B_1
Environmental Diagnostics Burlington, NC	EZ-Screen	Card	B_1, B_2, G_1, G_2
IDEXX Corp. Portland, ME	Probe	MTP	B_1, M_1
Int'l. Diagnostic Systems St. Joseph, MI	Afla-20	Cup	B_1, B_2, G_1
May and Baker (UK)	--	MTP, AC[2]	B_1, B_2, G_1
Neogen Corp. Lansing, MI	Agri-Screen	MTP	B_1
Oxoid Ltd. (UK)	--	AC	B_1, B_2, G_1, G_2
Penicillin Assays Inc. Malden, MA	CHARM	MTP	$B_1, B_2, G_1, G_2, M_1, M_2$
Transia Lyon, France	--	MTP, AC	B_1, B_2, G_1, M_1
UBE Tokyo, Japan	Afla-Check	MTP	B_1
Vicam Somerville, MA	Aflatest	AC	B_1, B_2, G_1, G_2, M_1

[1] microtiter well
[2] affinity column

Table II.   Agri-screen Collaborative Study - I
(14 Collaborators)

Commodity	Level* (ng/g)	ELISA Reader (ng/g)	RSD$_R$*** (%)	Visual** negative	positive
Corn	5	2.8±2.6	164		15
	26	12.8±8.1	73	36	
Peanuts,	6	21.8±7.9	31		85
raw	20	21.8±8.7	69	4	
Peanuts,	3	4.1±4.4	157		8
roasted	28	16.5±13.4	85	31	
Cottonseed,	36	35.2±15.9	89	17	
whole	85	41.0±12.3	42	0	
Cottonseed,	6	8.1±4.9	70		44
meal	31	18.6±9.3	51	12	
Mixed feed	4	5.9±5.3	93		35
	14	15.9±4.3	27	(23)	

*Determined using AOAC method 26.026-26.031 (CB) for corn
and peanuts, method 26.052-26.059 for cottonseed, and the
Shannon method for mixed feed (J.Assoc. Off. Anal. Chem.
1983, 66, 582.

**% false positive and negative results, i.e. for samples
containing <15 ppb a positive result using the kit would be
"false," for samples containing >15 ppb a negative result
would be "false."

***RSD$_R$: relative standard deviation between laboratories.

unacceptably high (>69%) for samples >15 ppb. There is no doubt that the AOAC's decision on the study occurred at a time when evaluation methodologies for such kits were in a very formative stage. ("Interim Official First Action" is the designation for a method whose performance characteristics have been evaluated by collaborative study and adopted by vote of the AOAC. After successful use for at least two years, the method may be adopted as "Official First Action".

Because of the less than satisfactory results obtained in the first study of the Agri-screen test kit when used with peanut products and corn, a second collaborative study was run. In this second study of the Neogen kit, twelve coded test samples of raw and roasted peanuts and corn, with blind replicates, were analyzed by fourteen laboratories. The determination was slightly modified in that tetramethylbenzidine was used instead of ABTS as substrate for color development, and the analysts were instructed to compare color intensity with standards of varying concentration when using visual estimation. The results of this study are shown in Table III (9).

Table III. Agri-screen Collaborative Study - II
(14 Collaborators)

Commodity	Level* (ng/g)	ELISA Reader (Ave. ng/g)	RSD_R*** (%)	Visual negative	positive
Corn	4.8	11.2	45.7		0
	31.0	56.0	52.7	0	
Peanuts,	1.5	9.2	43.5		6
roasted	39.0	70.2	23.3	0	
Peanuts,	10.8	125.0	62.2		13
raw	34.0	83.0	36.0	0	

*Determined using AOAC method 26.026-26.031 (Method 1) (14).
**%False positive and negative results, i.e. for samples containing <15 ppb a positive result using the kit would be "false," for samples containing >15 ppb a negative result would be false."
***RSD_R: relative standard deviation between laboratories.

Overall there was good correlation observed between ELISA and TLC results for corn and roasted peanut products, with 93 and 98% correct responses for visual and instrumental determinations, respectively. In the case of raw peanuts, a significant number of false positive results was noted for the low level sample (<20 ppb), as well as an extremely high instrumental result. However, it was subsequently determined that sample handling problems with raw peanuts were the source of this problem. This has been addressed, and another follow-up collaborative study planned. For visual determination in the <20 ng/g sample, the RSD_r (relative standard deviation within laboratory) and RSD_R (relative standard deviation between laboratories) for corn were 38.5 and 60.7% and for roasted peanuts 73.7 and 73.7%, respectively. These are considerably higher than the instrumental results. On the basis of this study the AOAC adopted the method as an Official First Action screening method for aflatoxin B, in corn and roasted peanuts at >20 ng/g.

A third collaborative study (10) was conducted examining the Immuno Dot Screen Cup (Afla-20, IDS Cup), a test kit containing antibodies with considerable cross reactivity with aflatoxins B_2 and G_1. The manufacturer's procedure was also modified to increase the reliability of detection at 20 ng/g total aflatoxins, and to broaden the applicability to include peanut butter samples.

In this study samples were sent to twelve collaborators. The peanut butter samples were blended with $MeOH/H_2O/hexane$ (55+45+100), filtered, the phases separated and the extract heated on a steam bath for two minutes to eliminate residual hexane; the corn, raw peanuts, cottonseed and poultry feed samples were blended with $MeOH/H_2O$ (80+20). After filtration, the extracts were diluted with buffer to <30% MeOH and aliquots applied to the kit cup in the bottom of which was a filter impregnated with the immobilized polyclonal antibodies. Aflatoxin B_1-peroxidase conjugate was then added followed by washing and then adding a mixture of H_2O_2 and tetramethylbenzidine. After exactly one minute, when no color was observed on the filter, the sample was judged to contain >20 ng/g aflatoxins; when a blue, or gray color developed, the sample was judged to contain <20 ng/g. The results of this study are tabulated in Table IV.

All collaborators correctly identified naturally contaminated corn (101 ng/g) and raw peanut (69 ng/g) positive samples. No false positives were found for control samples containing <2 ng/g. Overall (excluding poultry feed) the average correct responses for spiked positive samples at 10, 20 and >30 ng/g levels were 52, 86 and 96%, respectively (see Figure 2). The method was rapid and simple and was adopted Official First Action by the AOAC as a screening procedure for aflatoxins at $\geq$20 ng/g in cottonseed and peanut butter, and $\geq$30 ng/g in corn and raw peanuts. (In this case the AOAC required a positive rate of 90% for acceptance.)

Recently we completed an evaluation of the effectiveness of the Aflatest (Vicam) immunoaffinity column for analysis of corn, peanuts and peanut butter for aflatoxin (11). In this procedure the sample is extracted with $MeOH/H_2O$, filtered, and the extract diluted to <30% MeOH with water. An aliquot is then applied to the immunoaffinity column. The column is then washed with water and the aflatoxins eluted with methanol. Total aflatoxins are then determined by solution fluorometry with bromine (SFB), or individual toxins by reverse phase high performance liquid chromatography with post column I_2 derivatization (PCD). In the collaborative study corn samples naturally contaminated with aflatoxins and samples of corn, peanuts and peanut butter spiked at 30, 20 and 10 ng/g, were analyzed by 24 collaborators. The results of this study are tabulated in Table V.

From these data it appears that recoveries were consistently higher when measured using SFB, ranging from 97-131% for the three spikinglevels for the three commodities; recoveries by PCD ranged from 72-90%. Overall the interlaboratory variability (RSD_R) was found to be generally greater, by PCD. At the 20 ppb spiking level for the three commodities the RSD_R ranged from 14.4-20.6 using SFB, and

Table IV. IDS Cup Collaborative Study
(12 Collaborators)

Commodity	Level (ng/g)	Visual %+ve	95% Confidence Interval**	
			Lower	Upper
Corn	101 (NC*)	100	88	100
	30	92	73	99
	20	75	53	90
	10	33	16	55
	0	0	0	22
Peanuts, raw	69 (NC*)	100	88	100
	30	96	78	100
	20	83	62	95
	10	54	33	75
	0	0	0	22
Peanut butter	30	100	87	100
	20	91	72	99
	10	61	39	80
	0	0	0	24
Cottonseed	60	96	79	100
	20	96	79	100
	10	58	37	78
	0	0	0	22
Poultry feed	30	83	62	95
	20	46	26	67
	10	4	0	21
	0	0	0	22

*Determined using AOAC method 26.026-26.031 (14).
**Duplicate analyses for all spiked samples and naturally
  contaminated samples; single analysis for all control samples.
NC=Naturally Contaminated

Table V. Collaborative Study - Immunoaffinity Column
(24 Collaborators)

Commodity	Level*			Level Found			
		SFB	RSD$_R$	Recovery	PCD	RSD$_R$	Recovery
	(ng/g)	(ng/g)	(%)	(%)	(ng/g)	(%)	(%)
Corn	23(NC)	29	23.0	--	20	21.5	--
	30	33	20.0	109	24	11.7	83
	20	21	14.4	106	17	4.7	83
	10	12	33.1	124	7	50.8	72
	0	2	56.5	--	0.5	113.2	--
Peanuts	30	29	13.4	97	19	35.6	79
	20	21	15.3	105	16	16.2	80
	10	12	27.5	115	6	47.3	83
	0	0.8	107.2	--	0	268.5	--
Peanut	30	33	11.0	111	23	21.7	78
butter	20	21	20.5	107	16	23.1	81
	10	13	13.6	1?1	9	29.6	90
	0	2	36.5	--	0.4	123.1	--

*Determined using AOAC method 26.026-26.031 (CB).
NC = Naturally contaminated

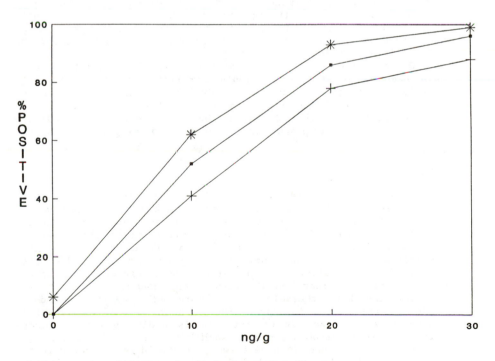

Figure 2. Operating characteristic curves for corn, peanut butter, raw peanuts, and cottonseed with 95% confidence limit.

4.63-23.1 using PCD. The data clearly indicate that both determinative procedures can be used with confidence even at 10 ppb.

Comparing these collaborative study results to the average expected from the Horwitz curve, a composite RSD_R curve of more than 300 collaborative studies. For SFB the average Horwitz ratios were 0.8, 0.59 and 0.55 for the three spiking levels; for PCD the ratios were 1.38, 0.33 and 0.96. Both SFB and PCD showed acceptable within and between laboratory precision. (The Horwitz ratio compares the RSD_R at the various levels and in the various matrices of this method with those RSD_R values predicted based historically on methods for a wide variety of analytes reported in AOAC collaborative studies; a ratio <2 is considered to have acceptable and typical precision (12).

Recently the USDA/FGIS conducted a collaborative study of six test kits for analysis of grains for aflatoxin (13). In the study three sample sets containing 21 ground corn samples each were analyzed in six laboratories. Two of the sample sets were spiked at 0, 10, 15, 20, 25, 30 and 40 ppb with 3 samples at each level; the third sample set contained naturally contaminated samples at the same approximate concentration levels. The results obtained using the kits were compared with those obtained using an official AOAC screening method, the Holaday-Velasco minicolumn method (14). The results are plotted as operating characteristics curves in Figure 3. Based on these results all of the kits, except the Agri-Screen kit, were found to have equivalent performance to the HV minicolumn. The FGIS subsequently reevaluated the Agri-Screen kit, which had been slightly modified by the manufacturer, and found the modified kit to be acceptable in screening corn for aflatoxin at the 20 ppb level in the USDA official inspection program (15).

Other collaborative studies of the commercially available kits have been reported. For example Mortimer and co-workers (16) recently reported on the use of the Quantitox B, ELISA kit produced by May and Baker Diagnostics, Glasgow in analysis of peanut butter. Their conclusion was that the kit can be used effectively to indicate aflatoxin B, levels in peanut butter, preferably using a peanut butter zero reference.

Conclusions

These studies demonstrate the excellent potential of immunoassay kits for use in screening commodities for aflatoxin. In reviewing the attempts made to date by many laboratories to accurately evaluate the qualities and capabilities of the various commercially available immunoassay test kits, it is clear that the manufacturers of the kits themselves have been constantly changing and improving the kits. This has hampered making value judgements and intercomparisons of the kits. It is also clear from the results obtained in collaboratively studying the kits themselves that more guidance is needed in thedesign, conduct and interpretation of the results of such studies. A generally accepted set of criteria is badly needed to ensure uniformity in intercomparisons of such kits. Finally, if a kit is designed to give a quantitative answer, it should be evaluated in the same manner AOAC has prescribed for other quantitative analytical methods (17).

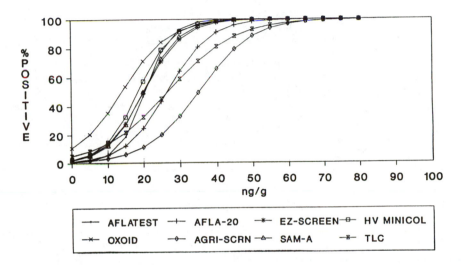

Figure 3.   Operating characteristic curves for USDA/FGIS collaborative study of aflatoxin test kits.

Literature Cited

1. Horwitz, W.; Cohen, S.; Hankin, L.; Keft, J.; Perrin C.H.;
    Thornburg, W. 1978, in Quality Assurance Practices for Health
    Laboratories, Inhorn, S.L.(Ed.), American Public Health Assoc.,
    Washington, DC, 588-589.
2. Food Chemical News, 1989, July 31, 47-48.
3. Food Chemical News, 1989, December 4, 7.
4. The Referee, 1989, 12 (11), 1-5.
5. J. Van Emon, Private Communication.
6. Koeltzow, D.E., Private Communication, July, 1989.
7. Federal Register, 1989, 54 (158), 33920-33923.
8. Park, D.L.; Miller, B.M.; Hart, L.P.; McVey, J.; Page, S.W.;
    Pestka, J.; Brown, L.H. J. Assoc. Off. Anal. Chem. 1989, 72,
    326-332.
9. Park, D.L.; Miller, B.M.; Nesheim, S.; Trucksess, M.W.;
    Vekich, A.; Bidigare, B.; McVey, J.; Brown, L.H. J. Assoc.
    Off. Anal. Chem. 1989, 72, 638-643.
10. Trucksess, M.W.; Stack, M.E.; Nesheim, S.; Park, D.L.;
    Pohland, A.E. J. Assoc. Off. Anal. Chem. 1989, 72, 957-962.
11. Trucksess, M.W.; Stack, M.E.; Nesheim, S.; Page, S.W.;
    Albert, R.H.; Hansen, T.J.; Kevin, F.D. J. Assoc. Off. Anal.
    Chem. 1990, in press.
12. Boyer, K.W.; Horwitz, W.; Albert, R. Anal. Chem. 1985, 57,
    454-459.
13. Koeltzow, D.E.; Tanner, S.N.; Private Communication, 1989.
14. Horwitz, W., AOAC Official Methods of Analysis 1984, Sec.
    26.020.
15. Koeltzow, D.E., Private Communication, August, 1989.
16. Mortimer, D.N.; Shepherd, M.J.; Gilbert, J.; Clark, C. Fd.
    Add. & Contam. 1988, 5, 601-608.
17. Horwitz, W.H. J. Assoc. Off. Anal. Chem. 1988, 71, 160-173.

RECEIVED May 8, 1990

Chapter 6

Immunochemical Assays

Development and Use by the California Department of Food and Agriculture

Peter J. Stoddard

Environmental Monitoring and Pest Management Branch, California Department of Food and Agriculture, Sacramento, CA 95814

Stimulated by legislative activity at the State level in 1986 concerning pesticide residues in ground water, CDFA began to develop immunochemical methods as both screening and quantitative assays, to reduce costs associated with an increased level of legally mandated environmental monitoring. So far, we have developed monoclonal antibodies for the triazine herbicides atrazine and simazine, and the rice herbicides molinate and thiobencarb, to be used in the ELISA "hapten tracer" format. Our experience so far suggests a need for a centralized location for information on and access to antibodies developed by other institutions worldwide.

The California Department of Food and Agriculture is testing immunochemical assays for possible adoption as analytical methods for pesticide residues in environmental samples. The work was initially undertaken by the Environmental Monitoring and Pest Management Branch (EM&PM), and now involves the Department's Chemistry Laboratory Services Branch as well. This paper discusses the history of our involvement in immunochemistry as an analytical method, our immunochemical method development program, the format and other details of the assay we are using, and finally, looks into the future in anticipation of some institutional problems which may arise.

History

Although the EM&PM Branch was interested in using immunochemical methods for analytical purposes as early as 1981, the program did not begin to move purposely toward adoption of methods until 1985 when the California Legislature began dealing with pesticide residues in ground water. The Pesticide Contamination Prevention Act, which became effective on January 1, 1986, and the Safe Drinking Water and Toxic Enforcement Act of 1986 which was approved by the voters in November of that year, both emphasized the need for less costly analytical methods to detect pesticide

0097–6156/90/0442–0051$06.00/0

residues in environmental samples. As a result of this heightened interest in ground water, the EM&PM Branch established contracts with Dr. Bruce Hammock at the University of California, Davis, Dr. Alex Karu, Director of the Hybridoma Center at the University of California, Berkeley, and Dr. Jeanette M. Van Emon of the Environmental Protection Agency's Environmental Monitoring Systems Laboratory in Las Vegas, Nevada. The express purpose of these contracts was to develop immunochemical assays for potential ground water leachers. The first of these assays have now been transferred to the Chemistry Laboratory Services Branch where they are being tested.

This origin in the EM&PM Branch resulted in two difficulties with which we have had to contend as we developed these assays. First, the pesticides which have proven to be leachers in California are organic chemicals of low molecular weight, and, in the case of the triazines, closely related structurally (Table I).

Table I. The seven pesticides for which residues have been found in ground water in California as a result of legal agricultural use, as of September, 1989

Common Name	Molecular Weight
Atrazine	216(a)
Simazine	202(a)
Prometon	226(a)
Diuron	233
Bromacil	261
Aldicarb	190
Bentazon	240

(a) These are triazines, which are closely related structurally.

These haptens have been coupled to carrier proteins in order to make them immunogenic. Their small size has limited the number of antigenic determinants, which in turn has made the isolation of antibodies and the design of analytical tests more difficult.   Thus, the Department has had to become familiar with immunoassays by working at the lower limit of the technology.

The second difficulty has resulted from the absence of regulatory levels for pesticides in ground water. Had the Department decided to introduce immunochemical assays first into the analysis of pesticide residues on raw agricultural commodities, we would have had clear guidelines for how much residue was tolerated (i.e., tolerances - see 40 CFR part 180), and we could have used the immunoassays to sort samples into two groups, one group over tolerance and the other under. However, perhaps because ground water is potentially vulnerable to so many different kinds of residues, there are no such levels established. The result is that if we want to use this technology as a screen (an assay used to sort

samples into two groups separated by a particular level of analyte),
the cutoff concentration for the screen is the minimum detectable
level for the conventional assay, usually gas or liquid chromatography. In our case, immunoassays have not had low enough minimum
detectable levels without a concentration step, which reduces the
simplicity of immunoassay.

Nevertheless, the Department has hopes that immunoassays will be
of increasing value as our experience with the technology increases.
In addition to detecting residues in environmental samples, it is
likely that immunoassays will eventually be used to detect residues
in raw agricultural commodities. Because of the tolerances established by the Environmental Protection Agency, immunoassays should
prove quite useful as a screening technique for these matrices.

Development and Use of Assays

To understand how the Department has selected target pesticides for
immunoassay development, it should be understood that the EM&PM
Branch monitors the environment only for residues of certain pesticides. Since we have regulatory authority only over those
pesticides in current use, we do not monitor the environment for
pesticides such as DDT and related chemicals, which are no longer in
use in the State. Therefore, although they are important environmental contaminants, they are not included in our assay development
program.

Of the registered pesticides in California, we have developed
antibody for four: the triazine herbicides atrazine and simazine,
and the rice herbicides thiobencarb and molinate. Of these, an assay for atrazine has been developed and is now being tested. It is
40% cross-reactive to simazine. Assays for the two rice herbicides
are now being developed.

We have divided our list of targets for future antibody production into two priorities listed in Tables II and III.

Table II. Pesticides being assessed for immunochemical assay of
soil and ground water samples by the California Department of
Food and Agriculture: First Priority

B. thuringiensis β-exotoxin
Bromacil CAS 314-40-9
Diuron CAS 330-54-1
Prometon CAS 1610-18-0
Ethylene thiourea CAS 96457
Fenamiphos (plus sulfoxide and sulfone) CAS 22224-92-6
Aldicarb (plus sulfoxide and sulfone) CAS 116-06-3
B. thuringiensis δ-endotoxin (excluding israelensis strain)
Diflubenzuron CAS 35367-38-5

Table III. Pesticides being assessed for immunochemical assay of soil and ground water samples by the California Department of Food and Agriculture: Second Priority

---

Alachlor CAS 15972608
B. thuringiensis δ-endotoxin (israelensis strain)
Carbaryl CAS 63-25-2
Captan CAS 133-06-2
Chlorothalonil CAS 1897-45-6
Cyanazine CAS 21725-46-2
2,4-D CAS 94-75-7
Dacthal CAS 1861-32-1
Diazinon (plus diazoxon) CAS 333-41-5
Endosulfan (α and β isomers and endosulfan sulfate) CAS 115-29-7
Ethylenebisdithiocarbamates: Maneb CAS 12427-38-2, Zineb
Glyphosate CAS 1071-83-6
Linuron CAS 330-55-2
Malathion (plus malaoxon) CAS 121-75-5
Methomyl CAS 16752-77-5
Metolachlor (Sencor) CAS 51218-45-2
Metribuzin CAS 21087-64-9
Paraquat CAS 1910-42-5
Pebulate (Tillam) CAS 1114-71-2
Propoxur (Baygon) CAS 114-26-1
Sulfometuron - methyl (no CAS number yet)
Triadimefon (Bayleton) CAS 43121-43-3
Tributyltin complex
Trifluralin (Treflan) CAS 1582-09-8

The Department plans to use immunoassays both as screening and quantitative analytical assays. They will be used by the Chemistry Laboratory Services Branch to analyze samples in the same way that Branch now uses the more traditional analytical methods. The extent to which immunoassays are integrated into normal use will depend on the results of tests which are now being designed and conducted in a joint effort between the two Branches.

In order to achieve maximum value to the Department, the immunoassays we develop will be validated through testing to determine within- and between-laboratory variability. At present, we plan to defer validation by the Association of Official Analytical Chemists (AOAC) until we have an assay which is more specific than the current atrazine assay, which is 40% cross-reactive to simazine.

Several ELISA formats have been considered, and the one currently being tested is called the "hapten tracer" method. First introduced to this project by Dr. Freya Jung while a postdoctoral fellow in Dr. Hammock's laboratory, this method uses microtiter plates coated with a commercial preparation of goat anti-mouse antibody. Enzyme-labeled hapten competes with analyte for receptor sites on a mouse monoclonal antibody specific for the analyte. The amount of enzyme left after washing is inversely proportional to the

amount of analyte in the sample. Other formats are also being examined and may be substituted or added as the need arises.

Much discussion has taken place in the past few years about the
desirability of monoclonal over polyclonal antibodies. We believe
each type of antibody has its unique uses, and discussion about the
relative merits of each should be referred to a specific context.
We have developed monoclonal antibodies initially because uniformity
over time was thought to be an advantage in a regulatory program.

Several field-portable kits are now available for detecting
chemical residues in the environment. The Department has obtained
and is testing these kits to determine whether they have a role in
our environmental monitoring program. However, the purpose of our
program is to develop highly precise, accurate and rapid immunoassays, uniform over time, yielding results which can withstand legal
challenges. In order to do this, we need a degree of control over
assay development and testing that is not available to us through
the use of commercial products.

Interest has been expressed in our antibodies by businesses exploring the possibility of using them for commercial purposes. As a
general principle, we support the transfer of this technology to the
private sector. However, all the reagents that have been produced
so far, including monoclonal and polyclonal antibodies and haptens,
are the property of the University of California and will be distributed by them. The Department has no plans to produce field-
portable kits at this time.

Problems to Overcome

As with any new technology, there will be obstacles to overcome in
the years ahead. Many of the technical obstacles have been pointed
out elsewhere in this symposium, and therefore we will focus attention on one problem which we have noticed as a regulatory agency,
and which is more institutional in nature.

As we develop the capability to use immunoassay, we are becoming
increasingly aware that antibodies and associated chemicals produced
elsewhere in the United States, and in foreign countries, would be
of value to us for internal regulatory use, and that our antibodies
would be of value to other states for the same purpose. However, we
have encountered a number of problems involving the dissemination of
this technology:

* It is difficult to keep abreast of the latest antibodies
developed. In particular, it is difficult to distinguish between
antibodies which have been developed, those which are in actual
preparation, and research developments regarding hapten linkages
which might produce effective antibodies in the future. Some important chemicals, such as ethylene thiourea, are a real challenge to
immunochemistry.

* Because the Department intends to rely on this technology for
regulatory purposes, the antibodies must be well characterized with
respect to specificity, cross reactivity and other parameters.
However, some individuals, wishing to protect their ability to
patent, license or market antibodies and related chemicals, are
reluctant to share information about the antibodies they have
developed, because they are unsure about how the information might
be used.

* Because the Department sometimes needs to develop assays rapidly in response to new environmental contaminants, it is necessary to obtain and store stocks of antibodies which may be needed in the future. Other states undoubtedly have the same need.

* There are no standards for transferring antibody and related chemicals between laboratories. This includes, among other things, standards for the amount of information about the antibody, and standards of purity and concentration.

* Patent attorneys who represent institutions which are developing antibodies and related chemicals, and who wish to retain rights to patent, license or commercialize those research products, approach negotiations defensively. It is easier to find out what they are not willing to do than to determine what they are willing to do. In California, we have found it very difficult to get patent attorneys to declare the nature of their interest and to explore ways the products of research can be used by public and private institutions.

The above problems suggest a need for a centralized location for:

* Information on what antibodies and related chemicals are being developed, which of these is ready to use, and training.

* Storage and rapid access to materials, and the development of standards for reagents submitted for storage.

* An agreement, centralized in some national or international organization, which would allow the information and materials to be used for internal regulatory purposes by states or other entities which agree to be a party to the agreement, while retaining patent, licensing and commercial rights for the originators of the technology.

The organization providing this coordinating function could be a governmental agency, such as the U.S. Environmental Protection Agency, or some other organization with perhaps a more international scope. However, the activities of such an organization should go beyond the simple transmission of information. This comment is meant to agree with and build upon a similar idea expressed by Dr. Hammock in his keynote address.

In closing, we would like to emphasize another comment made by the keynote speaker, that we should focus on a decade of success in immunochemistry. We have come a long way, both in technological advancement and in acceptance by the community of analytical chemists. Along the way we have had many individual successes and failures, but the momentum we have developed indicates that immunochemical assays for pesticide residues are here to stay.

Acknowledgments

The program described here has progressed as far as it has because of the involvement of many fine scientists, whose contributions we would like to acknowledge.

Work at the Hybridoma Center at the University of California, Berkeley, was directed by Dr. Alex Karu. Assisting him were Dr. Douglas J. Schmidt and Ms. Carolyn Clarkson.

Under Dr. Bruce Hammock's direction, the following people at the University of California, Davis contributed to the project: Dr. Robert Harrison, Dr. Marvin Goodrow, Dr. Freya Jung, and Ms. Shirley Gee.

Work at the U. S. Environmental Protection Agency's Environmental Monitoring Systems Laboratory, Exposure Assessment Research Division, was directed by Dr. Jeanette M. Van Emon. Also involved were Mr. Richard White (Lockheed - Engineering and Sciences Co.) and Mr. Kaz Lindley (University of Nevada, Las Vegas - Environmental Research Center).

Finally, at the Department of Food and Agriculture, the following people have also contributed significantly to the success of the program: Dr. Adolf Braun, Dr. Heinz Biermann, Ms. Catherine Cooper, Dr. Sylvia Richman, Mr. Vincent Quan and Ms. Jane Melvin.

RECEIVED May 22, 1990

Chapter 7

Immunoassay Methods

EPA Evaluations

Jeanette M. Van Emon

Environmental Monitoring Systems Laboratory, U.S. Environmental Protection Agency, Las Vegas, NV 89193-3478

The U.S. Environmental Protection Agency (EPA) has designated the Environmental Monitoring Systems Laboratory, Las Vegas, Nevada (EMSL-LV), as the EPA laboratory responsible for evaluating the application of immunochemical methods to environmental monitoring. To facilitate this designation, guidelines for evaluating immunoassay methods have been developed. These guidelines address criteria which should be met by the developer before an evaluation study is undertaken. The criteria addressed include: quality control; standard operating procedures; documentation of assay performance; data quality objectives; and supplies of specific antibody, antigen, and hapten. The degree of fulfillment of these criteria influence the design of the evaluation study. If only a few of the criteria are met, a preliminary evaluation will be conducted. These evaluation studies assist the developer to ultimately configure an assay appropriate for environmental monitoring. If the assay that is submitted to the EMSL-LV is well characterized, a single- or multi-laboratory evaluation may be conducted. For assays that are very mature and well-characterized, a short laboratory confirmation followed by an on-site field demonstration may be appropriate. Evaluation results are used to assess the application of immunochemical methods for environmental analysis.

Many EPA monitoring programs require analytical methods that are rapid and easy to perform on-site. Other monitoring programs, such as those for exposure assessment, require methods that can accommodate high sample loads with a rapid turnaround time. Immunoassay technology can provide both types of methodologies; however, the technology must be properly implemented in order to gain acceptance by the analytical community.

Because environmental immunoassays are a relatively new technology, evaluation studies are an important step to acceptance and implementation. Many environmental analytical chemists are not familiar with the technology, thus, any immunochemical methods that are introduced must be well-characterized and provide data of known quality. Another complication is that many environmental immunoassays are developed by individuals not trained in analytical chemistry. Evaluation studies are essential to provide appropriate quality assurance and quality control measures. Frequently, the analytical chemistries of extraction, sample cleanup, and recovery are not adequately addressed.

Evaluations will be conducted by the EMSL-LV only on those methods for which there is a perceived EPA application. Thus, evaluation studies are coordinated with monitoring activities where the immunoassay method could be implemented if deemed appropriate. This coordination also enables a better definition of needed performance characteristics to the targeted EPA user. Field portable methods may be evaluated under the EPA Superfund Innovative Technology Evaluation (SITE) program (1). The technology could then be used to monitor site containment integrity, remediation or cleanup activities as well as to monitor a site after remediation to ensure it remains free of contamination. The immunoassay may be appropriate for inclusion as an EPA method such as those described in "Test Methods for Evaluating Solid Waste, Physical/Chemical Methods" (2). At a minimum, this process requires method evaluation, publication in the Federal Register for public comment, review at EPA headquarters, and a cost analysis by the Office of Management and Business.

If an immunoassay technique is developed from outside the EPA, the cost of an evaluation is shared by the developer and the Agency, but the developer must provide all necessary reagents, supplies, and any specialized instrumentation. The EPA may provide personnel to conduct the evaluation, or volunteer laboratories may be used, for example, such as through the Association of Official Analytical Chemists (AOAC). The EPA is responsible for data analysis and report writing which includes a recommendation regarding the use of the particular technology. Evaluations are not undertaken as a service to the developer, but are performed to provide the EPA with new methods to fulfill analytical needs.

Evaluation Guidelines for Immunoassays

Ideally, before an evaluation study is undertaken, the developer should prepare a detailed oral presentation describing the development and application of the immunoassay. This will greatly assist in the planning of the evaluation study to ensure the proper experimental design as each evaluation is specific for each immunoassay. The proper experimental design is critical to fairly assess the method and not waste time and resources.

Evaluation, characterization, and testing of a particular analytical method is necessary to ensure the intended use of the method is met. In general, this process requires the determination of intra- and interlaboratory studies for precision and bias, method detection limits, matrix effects, interferences, limits of reliable measurements and ruggedness of the method. Before the EPA commits time and resources for an in-depth evaluation study, the developer must meet certain developmental criteria or justify why they were not met. The developer must also clearly define all necessary reagents as well as the underlying basis of the immunoassay.

Figure 1 presents the overall steps of the various EPA evaluations for immunoassays. Immunoassays are submitted to the EMSL-LV for evaluation. Raw data as well as data analyses must be submitted for review prior to initiation of a method evaluation. With the aid of a statistician, individual evaluation studies will then be designed to test the specific immunoassay and ensure that data quality objectives are met. These studies will be based on EPA and AOAC protocols for conducting evaluation studies (3).

Evaluation studies will be based primarily on synthetic samples (environmental samples spiked with known amounts of analytes). Samples should be selected which contain contaminants and interferences in concentrations likely to be encountered in actual real-world samples. Real-world samples (e.g., samples known to contain the contaminant of interest) will be used when available. When possible, the evaluation will be based on both synthetic (spiked) and real-world samples.

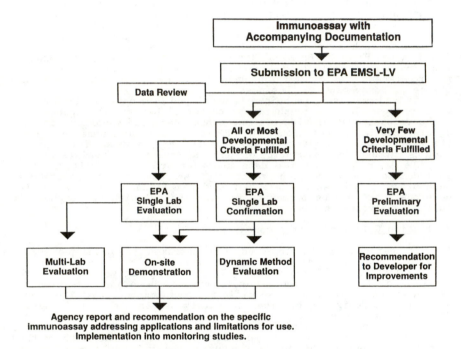

Figure 1.   The EMSL-LV immunoassay evaluation process.

The evaluation study will determine the attributes (bias, precision, specificity, limits of detection) of the immunoassay. Bias testing (systematic error) will be conducted by measuring recoveries of the analyte added to matrices of interest. Replicate analysis will be performed on blind replicates or split levels (e.g., Youden pairs). A minimum number of replicates will be performed to provide statistically meaningful results. The number of replicates will be determined by the intended purpose of the immunoassay as well as the documented method performance of the comparative method.

The immunoassay will be compared with an existing tested non-immunochemical analytical method, but this need not be a one-to-one evaluation performed on split samples. A sufficient number of samples will be split to be statistically meaningful. It is the intent of an evaluation to verify the statistical reliability of the immunoassay as an analytical method. The appropriate matrices to be analyzed will be determined by the data quality objectives. Analysis of the same material, on different days, will be performed by the immunoassay to provide information on temporal variability. Assay variability will be determined, for different days, with different calibration curves.

After successfully completing a single laboratory evaluation of a laboratory-based method, a multi-laboratory study may be undertaken. Laboratories participating in the multi-laboratory evaluation study must have personnel familiar with performing immunoassay methods. The number of laboratories participating in an interlaboratory evaluation will range between 3 and 6, depending on the scope of the project. The EMSL-LV will coordinate activities among the participating laboratories. The EPA may collaborate with the AOAC in multi-laboratory studies.

When appropriate, a dynamic (in-use) method evaluation of the assay will be performed following completion of the EMSL-LV single laboratory evaluation or confirmation. This type of evaluation is intended for immunoassays that are well-characterized and mature (i.e., a method where the developer has extensive performance data regarding matrix effects, cleanup, cross-reactivity, confirmatory analyses, and any other pertinent information). The data obtained during the dynamic evaluation will actually be used in a monitoring program. A dynamic evaluation can occur only where there is an immediate and urgent need for an analytical method.

Laboratory evaluations determine if a portable method is ready for field testing. If a method does not perform well in a laboratory evaluation it undoubtedly will fail in the field. If an assay is provided as a potential field screening method, then, after suitable laboratory evaluation, the assay will be taken to the field for an on-site demonstration. These evaluations will typically be conducted under the Agency's SITE program. The immunoassay technology is matched with an appropriate Superfund site and demonstrated, by the developer, on-site. Again, these types of evaluations are for those technologies which are ready for close scrutiny as the SITE program generates much publicity within and outside the Agency.

It is important that the evaluation of a field-portable assay method include performance data generated by personnel unskilled in immunoassay. The importance of an on-site demonstration is to test the method under actual field conditions. This ruggedness testing will determine the necessary precautions needed during field use. The precision, accuracy and bias obtained in the field will be compared to data obtained in the laboratory to fairly assess performance of the method obtained by unskilled personnel.

Whatever the format of the evaluation study, a final report is written which will contain the Agency's recommendations on the particular immunoassay and will address the applications and limitations of the method. Evaluation studies and final documentation will be completed as expeditiously as possible, however, there will be no compromise as to data quality.

Developmental Criteria

The immunoassay submitted for evaluation must be "mature," because expensive evaluation studies cannot be undertaken on unoptimized methods. The developer must first optimize and evaluate the immunoassay in-house, preferably by individuals most familiar with the assay system. Parameters addressed should include bias, precision (repeatability and reproducibility at the detection levels of interest), and rate of false positive and false negative results. The Agency should be supplied the raw data documenting assay performance. When possible, this information should be submitted on an IBM compatible floppy disk. A review of the data must be conducted before undertaking the evaluation study to verify interpretations of the data made by the developer.

The developer of a particular immunoassay must have well-defined data quality objectives to address the purpose of the immunoassay. These objectives should include the appropriate matrices to be analyzed, range of applicability for each matrix, specificity, detection limits, precision, and bias.

Each immunoassay must have a standard operating procedure (SOP). The SOP must address the entire analytical procedure (i.e., extraction, cleanup, detection, and data analysis and interpretation). Reagent stability and procedural precautions must also be included in the SOP.

Appropriate extraction protocols and sample preparation procedures must be well-documented by the developer and presented in a manner clearly understood by someone unfamiliar with the method. The effects of extraction and sample preparation on antibody performance must be determined. Specific areas such as pH, organic solvents, and ionic strength should also be addressed.

The effects of the appropriate environmental matrices (soil, water, air, biological - for biomarker or exposure assessment studies) on assay performance must be well characterized and documented. The SOP must also include the degree of quality control necessary to ensure the satisfactory performance of the method. Quality control procedures must address the required sample preparation steps, reagent stability, instrumentation, data handling and analysis. In many immunoassay SOPs that the EPA has reviewed, quality control is totally lacking or minimally addressed particularly for the sample preparations. The Agency can provide direction on what is an appropriate degree of quality control based on the objective of the method.

Even though a particular immunoassay is marketed as qualitative or semi-quantitative, the specific antibody must be characterized as quantitatively as possible with regards to cross-reactivity. Cross-reactivity studies must address: analogues, metabolites, degradation products, and non-target compounds which could be present in actual environmental samples (i.e., formulations components and mixtures of compounds for product efficacy). The Agency recognizes that it is not possible to test for all possible interferences. However, where feasible, efforts should be made to test the cross-reactivity of substances likely to be present in targeted real-world samples.

As the hapten is really the limiting reagent in antibody-based methods, the hapten synthesis and conjugation procedures must be documented and this documentation should be made available to the EPA as requested. All proprietary information will be protected with the appropriate procedures. However, the EPA should be convinced that any key reagent can be reproduced in a consistent manner. Ideally, the developer should stockpile "X" amount of hapten. The amount stockpiled should be based on an excess of 10-100 times the amount of assays expected to be performed in a reasonable time frame. These activities should help to allay fears of exhausting the supply of a limited reagent.

Immunization schedules and fusion protocols (for monoclonals) should be documented. A good antibody pool of either polyclonal or monoclonal must exist. Storage precautions, such as the use of multiple freezers, must be taken for the reagent antibody. Even the best monoclonal cell line is not "immortal" as monoclonals are ultimately dependent upon the proper storage.

When possible, the developer must provide confirmatory data on the immunoassay using another accepted analytical method (e.g., gas chromatography, gas chromatography/mass spectrometry, or high pressure liquid chromatography) that has been validated for the analyte and the matrix of interest. However, immunoassay can be applied to potentially hazardous substances which are not amenable to conventional methodologies.   Products of biotechnology such as those from genetically-engineered microorganisms may best be monitored by immunoassay. The EPA recognizes that a confirmatory methodology may not be available for the evaluation of these particular immunoassays.

Evaluation studies must be based on reagents from one specific antibody pool, one coating antigen pool, hapten enzyme conjugate pool or one pool of any key reagent. The developer must supply the Agency with the appropriate amount of immunologic reagents as well as all other materials necessary to complete the evaluation study. Other needed materials include items such as microtiter plates, antibody-coated test tubes, etc. Uniform lots of reagents and materials are critical for multi-laboratory evaluation studies. Thus, each laboratory participating in a multi-laboratory evaluation study must be supplied by the developer with identical immunologic reagents as well as other necessary materials.

Cost estimates regarding sample analysis as well as necessary accompanying instrumentation must be provided. The Agency acknowledges that some assays require specialized spectrophotometers/densitometers developed in conjunction with the assay; however, other tests are more flexible and are compatible with several different readers.

If unacceptable results are obtained following completion of an evaluation study, the developer must revise the method to fulfill the data quality objectives.   An alternative would be to state why the objectives could not be met and to revise the stated method performance specifications.

Summary

The evaluation process can be summarized by the following key milestones: 1) an identified EPA need, 2) submission of an immunoassay with accompanying documentation to the EMSL-LV, 3) data review, 4) Agency evaluation studies to provide additional performance data, 5) report addressing applications and limitations of the method, and 6) implementation of the immunoassay into routine monitoring programs.

It is the intent of these guidelines and developer criteria to provide clear definition of EPA evaluation protocols for immunoassays. The EPA is working with other agencies such as the U.S. Department of Agriculture, Food Safety and Inspection Services (4), and the Federal Drug Administration to develop mutually acceptable core evaluation requirements. It is anticipated that this coordination among agencies will assist developers and aid in the implementation of immunoassay into various governmental monitoring programs.

The EMSL-LV evaluates selected immunoassay methods to ensure the EPA has appropriate and necessary methods for environmental monitoring. When a developer approaches the EMSL-LV, efforts are made to find either an EPA Regional Office or a Superfund site where the immunoassay can be implemented for use.

NOTICE:

The information in this document has been funded (wholly or in part) by the United States Environmental Protection Agency. It has been subjected to Agency review and approved for publication.

Literature Cited

1. Comprehensive Environmental Response, Compensation, and
    Liability Act. Superfund, Section 311(b), Alternative or
    Innovative Treatment Technology Research and Demonstration
    Program.

2. Test Methods for Evaluating Solid Waste, Physical/Chemical
    Methods, 1-4, Third Edition, Office of Solid Waste Publication
    SW-846.

3. Horwitz, W., et al. Report of the Committee on Interlaboratory
    Studies. J. Assoc. Off. Anal. Chem., 1983, 66, 455-466.

4. Rapid Analytical, Diagnostic, and Microbiological Test
    Approaches, 54, USDA Federal Register, August 17, 1989,
    33920-33923.

RECEIVED August 1, 1990

ACADEMIC ADVANCES
IN IMMUNOASSAY TECHNOLOGY

Chapter 8

Polyclonal and Monoclonal Immunoassays for Picloram Detection

Raymond J. A. Deschamps and J. Christopher Hall

Department of Environmental Biology, University of Guelph, Guelph, Ontario N1G 2W1, Canada

A radioimmunoassay (RIA) and two indirect enzyme immunoassays for picloram (4-amino-3,5,6-trichloro-2-pyridinecarboxylic acid) were developed for the detection of the herbicide in river water, urine, as well as soil and plant extracts. The RIA method incorporated a rabbit anti-picloram serum as well as a novel radiolabel consisting of [^{3}H]glycine covalently linked to picloram. Using the RIA procedure, picloram concentrations in the range of 50 to 5,000 ng/ml could be detected in fortified river water and urine when a standard curve was prepared in the respective matrix. The I_{50} value for picloram by the RIA method was 760 ng/ml. The two indirect enzyme assays were compared in terms of sensitivity, accuracy and precision for detection of picloram in various fortified matrices using standard curves prepared in buffer. The assay using the same rabbit anti-picloram serum employed in the RIA method had a linear working range from 5 to 5000 ng/ml with an I_{50} value of 88 ng/ml and a lower detection limit of 5 ng/ml. The assay using a monoclonal antibody obtained from a mouse hybridoma cell line yielded a linear working range from 1 to 200 ng/ml with an I_{50} value of 10 ng/ml and a lower detection limit of 1 ng/ml. From the analysis of fortified river water, soil extracts, plant extracts, and urine, the monoclonal antibody-based enzyme immunoassay was shown to be more sensitive, more accurate, and more precise than the polyclonal antiserum-based enzyme immunoassay.

Picloram is used for the control of woody and broadleaf herbaceous plants. It is relatively resistant to breakdown in the environment and has been found to be mobile in the soil (1). Picloram residues

0097–6156/90/0442–0066$06.00/0

have been found in surface and groundwater samples (2, 3). The
mobility in the environment shown by picloram along with the
susceptibility of certain crops to extremely small amounts of this
compound (4) make monitoring water for picloram residues necessary.
     In disciplines such as clinical chemistry and endrocrinology,
immunochemistry is often the analytical method of choice because of
its sensitivity, specificity, speed of analysis, ease of automation,
cost effectiveness, and general applicability. The potential of
immunochemical technology for pesticide residue analysis in various
substrates such as soil, water, plants, urine, and blood has been
examined by several authors (5 - 11). Many pesticides, including
picloram, require an extensive sample preparation including
derivatization before they can be analyzed by gas chromatography. As
alternative methods, immunoassays can be sensitive, specific, and
precise providing for rapid, cost effective analyses. Immunoassays
may be based on polyclonal or monoclonal antibodies. The former is a
heterogenous mixture of proteins isolated from serum that represents
a variety of antibody molecules of differing specificities and
affinities. In contrast, the latter is a homogenous reagent
possessing a single antibody specificity and affinity. In a variety
of assay systems, either monoclonal or polyclonal antibodies may have
certain advantages over the other. For a detailed description and
comparison of polyclonal and monoclonal antibodies, the reader is
referred to the text by Zola (12).
     The majority of published immunoassay techniques for pesticide
detection employ polyclonal antibodies. In a review of immunoassays
for agrochemicals, Mumma and Brady (10) cite 49 assays employing
polyclonal antibodies and only 12 employing monoclonal antibodies.
The reason for this discrepancy in popularity may be that polyclonal
antibody-based assays, at first examination, are easier to develop.
However, when used as reagents for the quantitation of pesticide
residues, monoclonal antibodies have certain advantages over
polyclonal antibodies which include: i) hybrid cells can be cultured
indefinitely, either in vivo or in vitro to yield a potentially
unlimited supply of homogenous, standardized reagent; ii) during the
hybridoma selection process, the investigator can select a hybrid
cell producing the desired antibodies in terms of specificity and
affinity; iii) the monoclonal antibody will be free of antibodies
that are specific for irrelevant antigens which may interfere with
the assay's performance; and iv) cross-reactivity with structurally
similar molecules (e.g. other members of a pesticide class) can be
selected for or against depending upon whether the investigator
desires an assay to detect a single pesticide or a class of
pesticides (13, 14). Despite these issues which favor monoclonal
antibody-based assays, it is possible to develop excellent
immunoassays based on polyclonal antibodies.
     In this paper, we will review our previous research on a
radioimmunoassay (RIA) procedure for the detection and quantitation
of picloram using polyclonal antisera (15). Furthermore, we will
also discuss our research on indirect EIA procedures using monoclonal
and polyclonal anti-picloram antibodies which were compared in terms
of the characteristics of the standard curves and performance based
on the determination of picloram in fortified water, soil extracts,
plant extracts, and human urine samples (16).

Materials and Methods

Chemicals and Materials. The analytical standard of picloram clopyralid (3,6-dichloro-2-pyridinecarboxylic acid), fluroxypyr ([(4-amino-3,5-dichloro-6-fluoro-2-pyridinyl)oxy]acetic acid), triclopyr ([(3,5,6-trichloro-2-pyridinyl)oxy]acetic acid) along with radiolabelled picloram ([2,6-^{14}C]picloram, sp. act. 264 MBq/mmol) were provided by the Dow Chemical Company, Midland, MI. *N*-hydroxysuccinimide (NHS), *N,N'*-dicyclohexylcarbodiimide (DCC), isobutyl chloroformate, triethylamine, bovine serum albumin (BSA), rabbit serum albumin (RSA), Tween 20 (polyoxyethylene sorbitan monolaurate), Freund's complete adjuvant, and Freund's incomplete adjuvant were obtained from Sigma Chemical Company, St. Louis, MO. Aquasol 2 and [2-^{3}H]glycine (sp. act. 1609.5 GBq/mmol) were obtained from New England Nuclear Research Products, Boston, MA. Diethanolamine was obtained from Fisher Scientific Ltd., Don Mills, ON. Female Balb/cJ mice were obtained from The Jackson Laboratory, Bar Harbor, ME, or from Charles River Inc., Montreal, PQ. Cell culture media (RPMI and NCTC-109), fetal bovine serum, and the HAT selective medium components (hypothanthine, aminopterin, and thymidine) were obtained from Gibco Inc., Burlington, ON. Goat anti-rabbit IgG conjugated and goat anti-mouse IgG conjugated horse radish peroxidase were obtained from Jackson Immunoresearch Laboratories, Inc., West Grove, PA.

Preparation of Immunogens. Picloram was conjugated to BSA as described by Hall et al. (15) and Fleeker (17). Equimolar amounts of picloram (46 mg, with 45.5 Bq [^{14}C]picloram), NHS (22 mg), and DCC (39 mg) were dissolved in the sequence given in 2.5 mL of dioxane. The solution was allowed to stand at room temperature for approximately 18 h at which time it was filtered to remove the precipitate. The filtrate was evaporated to dryness on a rotary evaporator under vacuum at 35°C. A solution of BSA (500 mg) dissolved in 3 mL of 0.10 M borate buffer (pH 9, Fisher Scientific) was added to the residue and the mixture was agitated gently for 1 h at room temperature. The resulting solution was dialyzed against several changes of deionized water over 36 h at 4°C and lyophilized. The amount of herbicide bound to BSA was estimated by measuring ^{14}C present in weighed portions of product dissolved in PBS (0.01 M phosphate, 0.15 M NaCl). Approximately 15 to 20 molecules of picloram were bound per BSA molecule.

Preparation of Radiolabel. The mixed anhydride of picloram was prepared by adding picloram (6 mg), triethylamine (5 uL), and isobutyl chloroformate (5 uL) in the sequence given to 500 uL of dioxane. A portion of the mixed anhydride solution (100 uL) was added to a solution of 100 uL of [^{3}H]glycine (0.1 mCi), 100 uL of dioxane, 100 uL of distilled water, and 2 uL of 2 M NaOH. After 1 h, an additional 2 uL of NaOH was added. The reaction was allowed to proceed for a total of 4 h at room temperature.

The picloram-[^{3}H]glycine conjugate was isolated and purified by chromatography on silica gel TLC as described by Hall et al. (15) using the solvent system 60:40:2 diethyl ether: petroleum ether: formic acid (v/v/v).

Preparation of Coating Conjugates. Picloram (50 mg) was dissolved
in 5 mL thionyl chloride ($SOCl_2$) in a small boiling flask. The
solution was refluxed for 2.5 h at 85 °C to form the acid chloride of
picloram. Excess thionyl chloride was removed under vacuum at 60 °C
on a rotary evaporator. The residue was dissolved in 2 mL of
tetrahydrofuran (THF). The picloram acid chloride solution was added
slowly with stirring to 200 mg RSA in 10 mL of 0.02 N NaOH. Before
the addition of the acid chloride solution was completed, precipitate
formed which did not re-solubilize after stirring for 18 h at room
temperature. Dilution of the reaction mixture to 200 mLs with 0.02 N
NaOH succeeded in dissolving most of the precipitate. The resulting
suspension was centrifuged to remove any precipitate. The
supernatant was dialyzed against cold flowing tap water for 24 h and
lyophilized.

Production of Polyclonal Anti-Picloram Antibody. Picloram antisera
was obtained from New Zealand White rabbits following the protocol
described by Hall et al. (15). The rabbits were injected
subcutaneously with an emulsion consisting of 0.5 to 1.0 mg immunogen
dissolved in 0.5 mL of PBS and an equal volume of Freund's complete
adjuvant. The injections were repeated 3, 6, and 10 days after the
initial injection, substituting Freund's incomplete adjuvant for
complete adjuvant. A booster injection was given one month after the
initial injection and was repeated at monthly intervals thereafter.
The rabbits were bled for antibody titer determinations 10 days after
each boost. Antisera for picloram immunoassay development were
prepared from a single bleed in each case.

Production of Monoclonal Anti-Picloram Antibody. Ten 11-week-old
mice were injected intraperitoneally with a total volume of 250 uL of
a 1:1 (v/v) mixture of 70 ug of immunogen dissolved in PBS and
Freund's complete adjuvant (16). Secondary inoculations were given
three and eleven weeks after the initial immunization. One week
following each secondary inoculation, the mice were bled from the
retro-orbital plexus and the anti-picloram serum antibody titer was
determined using the RIA procedure described by Hall et al. (15). A
serum sample was considered positive for anti-picloram antibody
activity if binding of the picloram radiolabel was more than twice
the level of non-specific binding.
    The spleen was removed and cut into several small pieces and was
gently forced through a 400 mesh stainless steel screen into a Petri
dish which also contained RPMI medium. The cell suspension was
transferred to a sterile centrifuge tube and any large tissue
aggregates were removed by the sedimentation procedure described by
Shortman et al. (18). The suspension was centrifuged (200 x g) for
10 minutes and the cell pellet was resuspended in fresh medium.
Cells in trypan blue viability stain were enumerated microscopically.
The spleen cells were mixed with an equal number of SP/2.0 myeloma
cells in the semi-log growth phase in RPMI medium. The cell mixture
was centrifuged (200 x g) for 10 minutes and the cell pellet was
suspended in 1 mL of polyethylene glycol (3 000 - 4 000 mol. wt.
range) at 37 °C. The suspension was mixed continuously for one
minute, followed by the addition of 1 mL of RPMI medium and another
one minute of continuous mixing. An additional 9 mL of RPMI medium

was added slowly with mixing. The fusion products were centrifuged at 200 x g for 10 min, the supernatant discarded, and the cell pellet resuspended in RPMI medium supplemented with 10% fetal bovine serum, 10% NCTC-109 medium and 1% HAT. The cell suspension was dispensed (100 uL/well) into six sterile 96-well microtitration plates. The plates were incubated at 37 °C in an atmosphere of 5% CO_2 in air.

Fusion product culture supernatants were screened by enzyme immunoassay in a multi-stage manner to identify hybrid cell colonies producing anti-picloram antibodies. The various stages of the screening process were as follows: (i) identify wells containing colonies producing mouse IgG, (ii) indentify wells containing antibodies with activity against picloram-RSA coating conjugate, (iii) indentify wells containing antibodies specific for picloram as opposed to the carrier protein portion of the coating conjugate (RSA) by conducting a double screen using RSA and picloram-RSA coating conjugate, (iv) confirm specificity for picloram by attempting to inhibit binding of antibodies to picloram-RSA coating conjugate with picloram in solution, (v) identify wells containing antibodies specific for picloram as opposed to other pyridine herbicides by conducting cross-reactivity experiments with clopyralid, triclopyr, and fluroxypyr (Figure 1). Only the wells showing positive results were carried on to the next stage of screening. Out of 1152 cultures, only 5 remained at the completion of the screening process.

The culture showing the best results from an EIA assessment designed to detect the presence of picloram specific antibodies was selected for the limiting dilution procedure to acheive the clonality of the hybridoma cells. A dilution was calculated to yield one cell per well of a 96 well microtitration plate. The wells of the plate were checked daily for the presence of a single colony. Once a colony was visible, it was fed with 125 uL of RPMI medium. Supernatant (125 uL) was removed from the well for screening by EIA when the cells of the colony were one-quarter to one-half confluent. Cells from colonies testing positive for anti-picloram antibody activity were transfered to 24 well plates, rescreened by EIA and transfered again into 25 cm^2 flasks if they remained positive for picloram antibodies. The limiting dilution procedure was repeated to ensure monoclonality. After a final assessment by EIA, the cells producing the monoclonal antibodies specific for picloram were collected for the production of ascitic fluid in mice.

Mice were given an injection of 0.5 mL pristane (2,6,10,14-tetramethylpentadecane), a hybridoma growth promoting compound. Seven days later, the mice were injected with 3 x 10^6 hybridoma cells in 200 uL of PBS supplemented with 5% fetal bovine serum. Approximately two weeks following the injection of cells, ascites fluid was withdrawn, centrifuged to remove red blood cells, and frozen at -20 °C until used.

Sample Preparation. Water was collected from the Speed River, Guelph, Ontario and stored at 4 °C. The water was fortified with an acetone solution of picloram. Soil (40 g) was shaken 15 min with 200 mL of a 1:1 methanol/water solution. The mixture was filtered through a glass fibre filter and the methanol was removed under vacuum at 50 °C. The volume of the resulting aqueous solution was returned to 100.00 mL with Pi buffer (0.1 M phosphate, 1 mM $MgCl_2$, pH

7.5) and filtered through a 0.45 um nylon filter. The filtered
extract solution was fortified with an acetone solution of picloram.
Grass clippings (20 g) were homogenized in 100 mL of 0.1 N KOH with
10% KCl. The homogenate was shaken for 30 min and filtered through a
glass fibre filter. The filtrate was acidified to pH 2 with 3 N
H_2SO_4, refrigerated at 4 °C for 30 min and centrifuged at 3000 x g for
10 min. The volume of the supernatant was made up to 100.00 mL with
Pi buffer and aliquots were fortified with an acetone solution of
picloram. Prior to analysis, 10.00 mL of the fortified solution was
forced through a C_{18} reversed phase liquid chromatography column.
The column was washed with 5 mL of water and dried with a gentle
stream of forced air for 1 min. The column was eluted with 9 mL of
methanol. The eluate was evaporated to dryness and the residue was
redissolved in 10.00 mL of Pi buffer. Human urine was fortified with
an acetone solution of picloram. The urine was analyzed without
further processing by the RIA procedure. For the enzyme
immunoassays, 10.00 mL aliquots were acidified to pH 2 with 3 N
H_2SO_4. The picloram was extracted three times with 3 mL portions of
diethyl ether. The ether fractions were pooled and evaporated to
dryness. The residue was redissolved in 10.00 mL Pi buffer, and
cetrifuged at 12 000 x g for 10 min.

   Recoveries for the extractions described above were determined
using [^{14}C]picloram added to soil, grass clippings, and urine,
respectively. Recoveries were 95% for the soil extraction, 90% for
the plant extraction and 90% for the urine extraction.

Indirect Enzyme Immunoassay. The following procedure is a modified
version of that described by Deschamps et al. (16). Microtitration
plates were coated by adding to each well 200 uL of coating conjugate
dissolved in Pi buffer (0.1 ug coating antigen per mL). The plates
were incubated overnight at 4 °C. The plates were emptied and washed
three times with washing solution (Pi buffer supplemented with 0.1%
Tween 20). If the plates were not to be used immediately, they were
wrapped with plastic and and stored at 4 °C for up to 24 h.

   Sites on the polystyrene well surface unoccupied by coating
conjugate were blocked by adding 200 uL of 0.1% (w/v) gelatin
solution in Pi buffer and incubated for 20 min at 4 °C. The plates
were emptied and washed as described above.

   Antiserum diluted 1 to 20 000 or ascites fluid diluted 1 to 10
000 in Pi buffer supplemented with 0.05% Tween 20 surfactant (5) were
preincubated 1:1 (v/v) with picloram standard or sample solutions.
Aliquots of the preincubated mixture were transfered to the wells of
the microtitration plate (200 uL per well). The plates were
incubated for 1 h at 4 °C.

   After washing the plates as before, 200 uL of goat anti-rabbit
or goat anti-mouse IgG horseradish peroxidase conjugate diluted 1 to
5000 in Pi buffer was added to each well and the plates were
incubated for 1 h at 4 °C, emptied and washed.

   Substrate (1 mg/mL ABTS, 1 mg/mL urea hydrogen peroxide in
citrate buffer: 0.024 M citrate, 0.047 M phosphate, pH 5.0) was added
and color was allowed to develop for 30 min. The color reaction was
stopped by the addition of 100 uL 0.5 M citric acid. Absorbance of
each well was measured at 405 nm with a microtitre plate reader.
Absorbance values of the standards and the samples (A) were divided

by the maximum absorbance value (A_o) representing those wells in which binding of antibody to the coating conjugate was not challenged with free picloram in solution. The A/A_o values for standards were plotted against the log of picloram concentration to construct a standard curve. Concentrations of samples were calculated on the basis of the standard curve.

RIA Procedure. The RIA procedure described by Hall et al. (15) was used. Aliquots (100 uL) of standard or sample were transferred into 1.5 mL microcentrifuge tubes. Incubation mix (300 uL per tube) consisting of one part deionized water, one part inert serum, 12 parts PBS, and sufficient radiolabel to yield 10,000 cpm per assay was added to each tube. Antisera diluted in PBS was added to the tubes (100 uL per tube). One set of control tubes did not receive antisera for determination of non-specific binding and a second set of control tubes received antisera only for maximun binding of radiolabel (B_o). The contents of the tubes were mixed thoroughly, incubated at 4°C, the antibody-bound radiolabel fraction precipitated with $(NH_4)_2SO_4$, centrifuged, and the pellet dissolved in water prior to assaying for radioactivity.

All results were corrected for non-specific binding. Values for standards were divided by B_o and were plotted against the log of the herbicide concentration (ng/mL). The quantity of the herbicide in the unknown sample was calculated based on the standard curve.

Results and Discussion

Radioimmunoassay. A linear relation between the log of picloram concentration and relative binding (B/B_o) was found in the range of 50 to 5,000 ng/mL of picloram for the RIA procedure (Figure 2). The coefficient of variation within a run was 3% or less for picloram determined by the RIA method.

The accuracy of determinations of picloram in fortified river water and human urine samples determined by the RIA was good with the mean overall amounts detected varying from 82% to 110% of the amount of picloram added (Table I). The range of concentrations over which picloram was accurately quantitated with no sample clean-up correspond with levels found in urine in applicator exposure studies conducted by Libich et al. (19). It must be emphasized that determination of unknown concentrations of picloram in river water and urine was performed by using standard curves prepared in the respective matrix.

Table I. Picloram in Fortified River Water and Human
Urine Samples as Determined by RIA

Amount of picloram added, ug/mL	Amount of picloram determined[a]	
	River water	Human urine
0.25	0.25 ± 0.03 (6)	0.19 ± 0.04 (6)
2.50	2.60 ± 0.19 (6)	2.22 ± 0.35 (6)

[a]Mean amount determined: ug/mL ± SE (number of determinations).

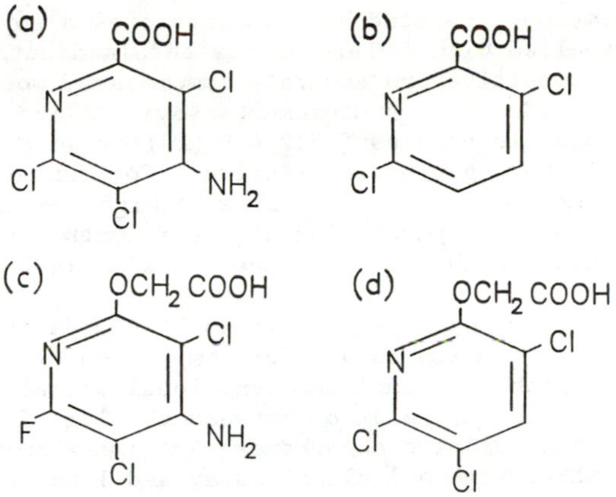

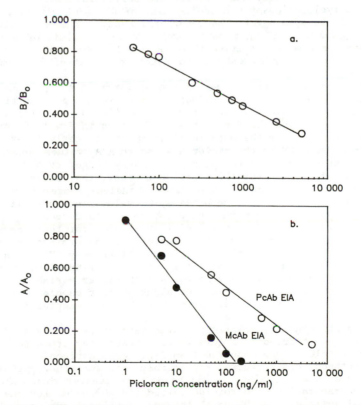

Figure 1. Chemical structures of (a) picloram, (b) clopyralid, (c) fluroxypyr, and (d) triclopyr.

Figure 2. Standard curves for: a. polyclonal antiserum-based radioimmunoassay and b. polyclonal antiserum-based and monoclonal antibody-based enzyme immunoassays (PcAb EIA and McAb EIA, respectively).

The RIA method reported here incorporates a novel radiolabel. Herbicides labelled with ^{14}C are easily obtained but do not lend themselves to sensitive and accurate immunoassay work because of low specific activities (8). Radioimmunoassays utilizing high specific activity radiolabels such as [3H]2,4-D (20) or an [125I]2,4-D derivative (21) have given good results. Covalently linking picloram with [3H]glycine yields a radiolabel with high specific activity without the expense of purchasing a custom synthesized tritiated herbicide or the health hazards connected with iodated radiolabels.

Indirect Enzyme Immunoassays. Picloram standards in Pi buffer were used to generate standard curves for comparison of two indirect EIA procedures in which polyconal and monoclonal antibodies were used, respectively. A linear relation between the log of picloram concentration and relative absorbance (A/A_o) was found in the range 5 to 5000 ng/mL for the polyclonal assay and 1 to 200 ng/mL for the monoclonal assay (Figure 2). The monoclonal assay, therefore, had a standard curve with a much steeper slope compared to the polyclonal assay. Typical coefficient of determination values (r^2) were 0.97 for the monoclonal assay and 0.95 for the polyclonal assay.

The polyclonal assay had an I_{50} value of 88 ng/mL with a lower detection limit of 5 ng/mL. The monoclonal assay was more sensitive with an I_{50} value of 10 ng/mL and a lower detection limit of 1 ng/mL. Both assays were more sensitive than the RIA for picloram which had an I_{50} value of 760 ng/mL and a lower detection limit of 50 ng/mL (15).

Using the absorbance values of the picloram standards, the interwell variability was determined for the two EIA procedures. The polyclonal assay showed a mean interwell coefficient of variation (CV) of 6.4% over the standard curve. The mean interwell CV over the standard curve for the monoclonal assay was slightly lower at 5.3%. Interassay CV of the picloram standard A/A_o values determined on four separate runs for the polyclonal assay ranged from 2.1 to 23% with a mean of 12.8%. For the monoclonal assay, the interassay CV of A/A_o values determined on seven separate occasions ranged from 5.1 to 26% with a mean of 16%. In both cases, CV values increased with an increase in picloram standard concentration due to decreasing A/A_o values. Singh et al. (22) showed similar results for their enzyme immunoassay intended for routine slaughterhouse determinations of the antibiotic sulfamethazine in swine plasma. Intrassay CV values were obtained on picloram determinations in four fortified plant extract samples. The polyclonal assay showed a much higher variability with a mean CV value of 80% over the four plant extract samples compared to only 19% for the monoclonal assay over the same samples (Table II).

Three structurally related pyridine herbicides, clopyralid, fluroxypyr, and triclopyr (Figure 1) were tested for cross-reactivity with the polyclonal and monoclonal anti-picloram antibodies. Neither antibody cross-reacted appreciably with the other pyridine herbicides as the I_{50} values in all cases were greater than the highest concentration of herbicide tested (50 000 ng/mL for the polyclonal antibody, 10 000 ng/mL for the monoclonal antibody).

Determinations of picloram in fortified water, soil extracts, plant extracts, and urine indicated that the monoclonal assay was far superior for quantitative determinations (Tables III, IV).

Table II. Intraassay Variability of Picloram in Four Fortified Plant
Extract Samples from Enzyme Immunoassay Standard Curve using
Polyclonal or Monoclonal Antibodies

| Picloram added, ng/mL | Picloram determined | | | |
	PcAb EIA[a] mean, ng/mL	CV, %	McAb EIA[b] mean, ng/mL	CV, %
0	3.6	65	1.1	32
4	9.9	87	3.9	29
20	39	89	24	15
40	99	84	52	21
400	780	59	450	10

[a]Polyclonal antibody enzyme immunoassay.
[b]Monoclonal antibody enzyme immunoassay.

Table III. Recovery of Picloram from Fortified Water, Soil and Plant
Samples Determined by Enzyme Immunoassay using Polyclonal or
Monoclonal Antibodies

| Picloram added, ng/mL | Picloram determined, ng/mL[a] | |
	PcAb EIA[b]	McAb EIA[c]
Fortified water		
20	27 ± 5.1 (18)	11 ± 0.98 (12)
200	569 ± 79 (18)	165 ± 7.4 (12)
2000	3590 ± 550 (18)	1920 ± 80 (12)
Fortified soil		
4	23 ± 4.4 (24)	2.1 ± 0.15 (36)
20	90 ± 23 (24)	13 ± 0.53 (36)
40	110 ± 23 (24)	33 ± 1.1 (36)
400	1010 ± 280 (24)	480 ± 13 (12)
Fortified plant extract		
4	9.9 ± 2.5 (12)	3.5 ± 0.23 (24)
20	39 ± 10 (12)	24 ± 1.0 (24)
40	99 ± 24 (12)	51 ± 2.4 (24)
400	780 ± 130 (12)	450 ± 13 (24)

[a]Mean ± SE (number of determinations).
[b]Polyclonal antibody enzyme immunoassay.
[c]Monoclonal antibody enzyme immunoassay.

Table IV. Recovery of Picloram from Fortified Human Urine Samples
Determined by Enzyme Immunoassay using a Monoclonal Antibody[a]

Picloram added, ng/mL	Picloram recovered, ng/mL[b] McAb EIA[c]
4	11 ± 0.58 (12)
20	30 ± 2.2 (12)
40	50 ± 2.1 (12)
400	450 ± 13 (12)

[a]Polyclonal assay was not successful due to unknown contaminant.
[b]Mean ± SE (number of determinations).
[c]Monoclonal antibody enzyme immunoassay.

Overall mean values of picloram determined for the monoclonal assay
were 78, 73, 112, and 167% of the amount added for water, soil
extract, plant extract, and urine, respectively. For the polyclonal
assay, overall mean values of picloram determined were 200, 388, and
221% of the amount added for water, soil extract, and plant extract.
The polyclonal assay for determination of picloram in urine was not
successful because of unacceptable interference from an unknown
contaminant. Picloram concentration estimates were taken from a
standard curve made in Pi buffer. Interference from components of
the sample matrix likely accounts for much of the error in the
concentration estimates. Such interferences from sample components
have been reported by Wie and Hammock (23). In preliminary studies,
we have found that the ionic strength of the matrix solution and
possibly the presence of organic co-extractives influence the amount
of sample matrix interference. These effects were minimized if the
antibodies were diluted in Pi buffer supplemented with 0.05% (v/v)
Tween 20 as described by Hunter and Lenz (9). The RIA procedure for
picloram in water and urine using the same polyclonal antisera as in
the EIA described here showed a high degree of accuracy (82 to 110%
recovery) when the standard curves were constructed using blank water
or urine. Singh et al. (22) used swine plasma as the reference
matrix in their EIA to determine the antibiotic sulfamethazine in a
swine plasma sample matrix with excellent accuracy. We chose not to
do this for the indirect monoclonal and polyclonal procedures for the
following reason. Soil or water samples from different geographical
regions or urine samples from different subjects will vary widely in
composition. Selecting one blank water or soil sample to use for the
standard curve would not be appropriate and attempts to obtain a
representative water or soil sample would be difficult. Upon
examination of the determinations of picloram in the various sample
matrices (Tables III and IV), it is evident that this choice had more
severe consequences with respect to the polyclonal antibody-based
assay than with the monoclonal antibody-based assay. The advantage
of a standard curve with a steep slope is that small errors in
absorbance values will not translate to large errors in concentration
estimates.
        One disadvantage of a standard curve with a steep slope is the
narrow linear working range. Rather than making several dilutions of
a sample in the hope of obtaining one dilution in the proper range,

it may be more efficient to conduct a separate assay with a wide
working range to rank samples so that appropriate dilutions can be
made with certainty for accurate quantitation by a second assay. The
polyclonal system described here would be adequate for the role of
ranking samples. Alternatively, one could modify the parameters of
the monoclonal assay (e.g., increase the antibody concentration) to
achieve a standard curve with a flatter slope and a wider working
range.

For polyclonal antibody production, the design and the
preparation of the immunogen are most critical (7, 8, 24). Several
studies have illustrated the influence of hapten structure, bridging
groups, immunogen structure, and coating conjugate structure on
immunoassay performance (25 - 27). The goal of immunogen design and
preparation is to maximize the quantity of specific antibodies in the
antisera having high affinity for the antigen (analyte). Although it
can not be denied that the development of a good immunogen at the
outset is the most effective way to obtain a good antibody, the
design of the immunogen used to produce monoclonal antibodies may not
be as critical as that required for polyclonal antibody production.
An effective screening program will enable the investigator to select
and expand the hybridoma cell clone(s) producing the desired antibody
even if such clones are rare. In the present study, the same
immunogen that yielded a polyclonal antisera with a low average
affinity also yielded a monoclonal antibody of high affinity based on
the I_{50} values reported here.

These immunoassays could be incorporated on a routine basis in
most laboratories to serve one of two functions. The assays could
be used as a rapid, inexpensive method for herbicide quantitation
with little or no sample clean-up. Alternatively, they may be
implemented as a preliminary screen to rank samples for follow-up
determination by gas chromatography. In either function, the
immunoassays represent savings in time, labor, and materials.

Acknowledgments

The provision of radiolabeled [^{14}C]picloram and analytical picloram,
fluroxypyr, clopyralid, and triclopyr by the Dow Chemical Company are
gratefully acknowledged. Suppport for R.J.A.D. was provided by a
Natural Sciences and Engineering Research Council of Canada Post-
Graduate Scholarship. The research was supported by grants to J.C.H.
from the Natural Science and Engineering Research Council of Canada,
the Ontario Ministry of the Environment, The Ontario Pesticide
Advisory Committee, the Ontario Minsitry of Agriculture and Food, as
well as Agriculture Canada.

Literature Cited

1. Hamaker, J. W.; Johnston, H.; Martin, R. T.; Redemann, C.
    T. Science 1963, 141, 363.
2. Frank, B.; Clegg, B. S.; Ripley, B. D.; Braun, H. E. Arch.
    Environ. Contam. Toxicol. 1987, 16, 9-22.
3. Baur, J. R.; Bovey, R. W.; Merkle, M. G. Weed Sci. 1972,
    20, 309-313.
4. Ragab, M. T. H. Can. J. Soil Sci. 1975, 55, 55-59.
5. Cheung, P. Y. K.; Gee, S. J.; Hammock, B. D. In Impact of
    Chemistry on Biotechnology; Phillips, M., Shoemaker, S.

P., Middlekauf, R. D., Ottenbrite, R. M., Eds.; ACS
Symposium Series 362; American Chemical Society:
Washington, DC, 1988; pp 217-229.

6.   Ercegovich, C. D.; Vallejo, R. P.; Gettig, R. R.; Woods,
L.; Bogus, E. R.; Mumma, R. O. J. Agric. F. Chem. 1981, 29,
559-563.

7.   Hammock, B. D.; Gee, S. J.; Cheung, P. Y. K.; Miyamato, T.;
Goodrow, M. H.; Van Emon, J. Seiber, J. N. In Pesticide
Science and Biotechnology; Greenhalgh, R.; Roberts, T. R.,
Eds.; Blackwell Scientific Publications: Oxford, 1987, pp
309-316.

8.   Hammock, B. D.; Mumma, R. O. In Recent Advances in
Pesticide Analytical Methodology; Harvey, J. Jr.; Zweig,
G., Eds.; ACS Symposium series 136; American Chemical
Society: Washington, DC, 1980; pp 321-352.

9.   Hunter, K. W., Jr.; Lenz, D. E. Life Sci. 1982, 30, 355-361.

10. Mumma, R. O.; Brady, J. F. In Pesticide Science and
Biotechnology; Greenhalgh, R.; Roberts, T. R., Eds.,
Blackwell Scientific Publications: Oxford, 1987; pp 341-348.

11. Newsome, W. H.; Shields, J. B. J. Agric. F. Chem. 1981, 29,
220-222.

12. Zola, H. Monoclonal Antibodies: A Manual of Techniques; CRC
Press: Boca Raton, 1987; p 214.

13. Vanderlaan, M.; Van Emon, J.; Stanker, L. In Pesticide Science
and Biotechnology; Greenhalgh, R.; Rooberts, T. R., Eds.;
Blackwell Scientific Publications: Oxford, 1987; pp 597-602.

14. Vinas, J. Pure Appl. Chem. 1985, 57, 577-582.

15. Hall, J. C.; Deschamps, R. J. A.; Krieg, K. K. J. Agric.
Food Chem. 1989, 37, 981-984.

16. Deschamps, R. J. A.; Hall, J. C.; McDermott, M. R. J.
Agric. Food Chem. 1989, (submitted).

17. Fleeker, J. R. J. Assoc. Off. Anal. Chem. 1987, 70, 874-
878.

18. Shortman, K. N., Williams, N., Adams, T. J. Immunologic
Meth. 1972, 1, 273-279.

19. Libich, S.; To, J. C.; Frank, R.; Sirons, G. J. Am. Ind.
Hyg. Assoc. J. 1984, 45, 56-62.

20. Knopp, D.; Nuhn, P.; Dobberkau, H.-J. Arch. Toxic. 1985,
58, 27-32.

21. Rinder, D. F.; Fleeker, J. R. Bull. Environ. Contam.
Toxicol. 1981, 26, 375-380.

22. Singh, P.; Ram, B. P.; Sharkov, N. J. Agric. Food Chem.
1989, 37, 109-114.

23. Wie, S. I.; Hammock, B. D. J. Agric. Food Chem. 1982, 30,
949-957.

24. Jung, F.; Gee, S. J.; Harrison, R. O.; Goodrow, M. H.;
Karu, A. E.; Braun, A. L.; Li, Q. X.; Hammock, B. D.
Pestic. Sci. 1989, 26, 303-317.

25. Vallejo, R. P.; Bogus, E. R.; Mumma, R. O. J. Agric. Food
Chem. 1982, 30, 572-580.

26. Wie, S. I.; Hammock, B. D. J. Agric. Food Chem. 1984, 32,
1294-1301.

27. Wie, S. I.; Sylwester, A. P.; Wing, K. D.; Hammock, B. D.
J. Agric. Food Chem. 1982, 30, 943-948.

RECEIVED May 8, 1990

Chapter 9

Trinitrotoluene and Other Nitroaromatic Compounds

Immunoassay Methods

D. L. Eck, M. J. Kurth, and C. Macmillan

Department of Chemistry, Sonoma State University,
Rohnert Park, CA 94928
Department of Chemistry, University of California, Davis, CA 95616

Several enzyme-linked immunosorbent assays (ELISAS) have been developed for trinitrotoluene, trinitrobenzene, 2,4-dinitrotoluene, and 2,6-dinitrotoluene using polyclonal antibodies raised in New Zealand white rabbits. Nitro substituted benzoic and phenyl acetic acids were used as haptens by conversion to the correspond NHS esters followed by coupling to protein carriers.The antibodies which were developed to 1,3-dinitroaromatic haptens had the greatest specificity and sensitivity when the nitroaromatic analytes contained a 1,3-dinitro functionality. In one ELISA system a lower detection limit for various 1,3-dinitroaromatics analytes of 1 ng/mL with an I_{50} of 5 ng/mL was observed. No cross reactivity with mononitroaromatic compounds was observed. Antibodies developed to mononitroaromatic haptens showed high affinity for a variety of coating antigens but would not compete with nitroaromatic analytes in a normal ELISA.

The ever increasing production of synthetic organic chemicals coupled with a growing public awareness and concern over the environmental fate and safety of these substances has made it increasingly important for scientists to provide our highly technological society with means to inexpensively and accurately monitor chemical substances in the environment. Information about the total amount of a chemical residue, whether it be in a hazardous materials site, a food crop or human subject is important in determining the relative risk that substance poses to the environment. A limiting factor in many conventional analytical methods of monitoring environmental pollutants has been the highly technical and costly equipment required for the successful analysis of these chemical substances. In addition, the high cost per individual analysis often limits the number and frequency of tests and, since the equipment is not portable, scientists are often limited in their ability to do timely tests in the field.

Immunochemical methods are rapidly gaining acceptance as an option for residue and other environmental analyses. They offer the advantages of being reliable, simple, relatively inexpensive, and field adaptable alternatives to

0097–6156/90/0442–0079$06.00/0

conventional chromatographic and colorimetric methods. Several excellent reviews have outlined the merits as well as drawbacks of immunoassays and have described the overall methodology involved in developing immunochemical assays (1-4). The technique has been successfully used to assay a variety of chemical substances. For example the fungicide fenpropimorph (5), the herbicides 2,4-D (6), picrolam (6), and molinate (7) and the insecticide diflurobenzuron (8) have all been successfully analyzed in environmental samples.

A likely set of compounds to consider for similar assay development are the nitroaromatic compounds shown in Figure 1 which are related to nitrobenzene (1). As a class, nitroaromatic compounds are of environmental concern since, as Figure 1 shows, they have been documented at as many as 30 of the 818 final EPA National Priority List (NPL) of waste sites in the United States (9). The nitroaromatics most frequently found as environmental contaminants are 2,4-dinitrotoluene (2) and 2,6-dinitrotoluene (3) used in plastics, dyes and munitions production; nitrophenols (4,5) used in pesticides; and munitions wastes such at 2,4,6-trinitrotoluene (6) or 1,3,5-trinitrobenzene (7).

Large quantities of nitroaromatic compounds are currently manufactured in the United States. For example, the 1982 production of dinitrotoluenes exceeded 720 millions pounds (10) while the 1985 production of nitrobenzene exceeded 914 million pounds (11). The toxicity, mutagenicity, and carcinogenicity of nitroaromatics such as 2,4- dinitrotoluene are well established and have been the subject of numerous reports and reviews (12-15). Therefore, the need for extensive monitoring of nitroaromatics in the environment (production effluents, toxic waste disposal sites, work places, etc.) clearly exists. The well documented immunogenic response of the nitro functional group in other aromatic compounds is an additional reason for selecting the targets in Figure 1 for immunochemical analysis (16).

Synthesis Strategy

The nitroaromatic compounds of interest are too small to illicit an immune response in test animals. Therefore, as in any small molecule immunoassay, a key element is hapten-protein conjugate design particularly as one hapten-protein conjugate is needed for rabbit immunization and a second hapten-protein conjugate is needed as a microtiter plate coating antigen. Haptens selected for this investigation were the nitro substituted benzoic acids and phenyl acetic acids shown in Figure 2. The nitroaromatic portion of these molecules was expected to trigger the immune response while the carboxylic acid portion when activated with a N-hydroxysuccinimide leaving group(NHS) (8a-14a) would serve as the linker arm to conjugate the nitroaromatic haptens to a lysine unit in the protein. In this study haptens conjugated to bovine serum albumin(BSA)(8b-14b) were used as immunizing antigens and hapten conjugated to chicken egg albumin (OVA)(8c-14c) were used as coating antigens.

Four structures (8b-11b) were selected to be injected into rabbits as immunizing antigens, their relationship to the target analytes in Figure 1 being obvious. For example, the BSA conjugate of 2,4-dinitrophenylacetic acid (8b) seemed a likely immunogen for 2,4-dinitrotoluene or trinitrotoluene while the BSA conjugate of 3,5-dinitro-4-methylbenzoic acid (11b) was selected as a target for 2,6-dinitrotoluene. The BSA conjugates of 4-nitrophenylacetic acid (9b) and

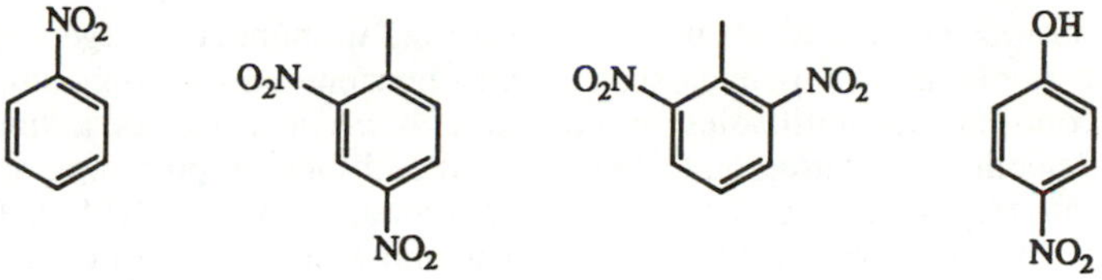

1 (2)[a]   2 (9)[a]   3 (8)[a]   4 (3)[a]

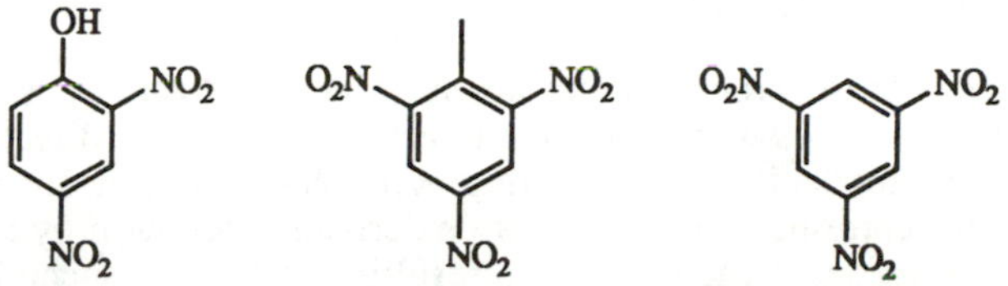

5 (2)[a]   6 (13)[a]   7 (4)[a]

[a] The number of NPL sites contaminated with this compound (9).

Figure 1. Structures and frequency of occurrence of nitroaromatic compounds at U.S. EPA National Priority List sites.

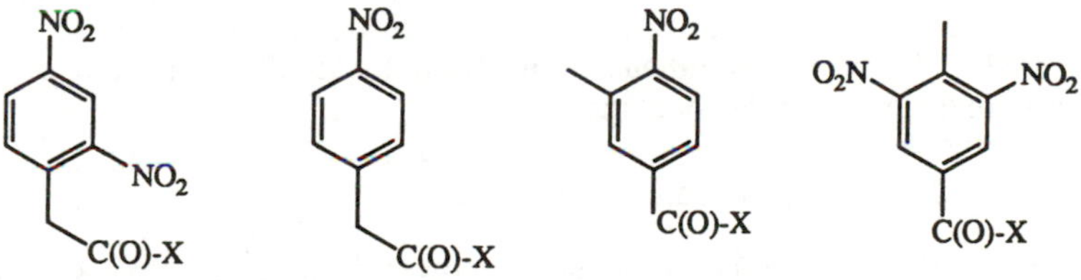

8a X = NHS   9a X = NHS   10a X = NHS   11a X = NHS
8b X = BSA   9b X = BSA   10b X = BSA   11b X = BSA
8c X = OVA   9c X = OVA   10c X = OVA   11c X = OVA

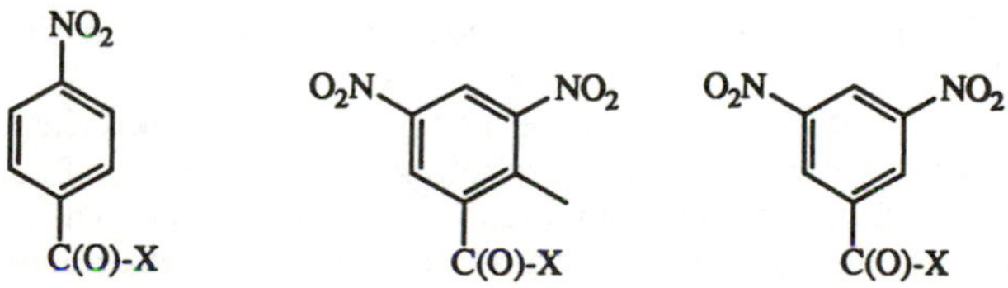

12a X = NHS   13a X = NHS   14a X = NHS
12b X = BSA   13b X = BSA   14b X = BSA
12c X = OVA   13c X = OVA   14c X = OVA

Figure 2. Structures of the N-hydroxy succinimide (NHS) esters of selected nitroaromatic haptens, the hapten-bovine serum albumin (BSA) conjugates and the hapten-ovalbumin (OVA) conjugates.

3-methyl-4-nitrobenzoic acid (**10b**) were chosen as immunogenic agents for the detection of mononitro compounds such as nitrobenzene and nitrophenol (**4**). It was also proposed that antibodies to **9b** and **10b** might serve as a "universal indicator", detecting any nitroaromatic compound and thereby providing a system capable of broadly screening for all of the compounds on the NPL list. The antibodies to **8b** and **11b** could then be used to more specifically determine the identity of the compound.

Synthesis of Protein Conjugates:

The synthesis of the protein conjugates is shown in Figure 3. Commercially available carboxylic acids were converted to activated esters by direct DCC (N,N-dicyclohexylcarbodiimide) coupling with N-hydroxysuccinimide or by conversion to the corresponding acid chloride derivative followed by reaction with N-hydroxysuccinimide. Reactiion of the resulting N-hydroxysuccinimide esters with either BSA or OVA led to the desired lysine bonded nitroaromatic hapten-protein conjugates.

Reagents:

Nitro substituted carboxylic acids, N,N-dicyclohexylcarbodiimide (DCC) and N-hydroxysuccinimide were obtained from the Aldrich Chemical Co. (Milwaukee, WI). Avidin-labeled horseradish peroxidase, biotinylated anti-rabbit IgG, ovalbumin (M.W. 45,000), bovine serum albumin (M.W. 66,000) , Freund's complete adjuvant, Freund's incomplete adjuvant, and *o*-phenylenediamine were purchased from Sigma Chemical Co.(St. Louis, MO). All solvents were reagent grade. DME (1,2-dimethoxyethane) was dried by distillation from sodium-potassium amalgam/benzophenone ketyl.

Preparation of Active N-Hydroxysuccinimide Esters

Method A. A stirred solution of 4 mmoles of nitro-substituted carboxylic acid and 4 mmoles of N-hydroxysuccinimide in 20 mL of dry DME was cooled in an ice bath. A solution of 4.4 mmole of DCC in 6 mL of dry DME was added dropwise (30 min). The resulting mixture was stored in a refrigerator overnight. The cold solution was filtered to remove the dicyclohexylurea precipitate. The precipitate was washed with 10 mL of dry DME. Removal of the solvent in vacuo yielded the crude NHS ester.
Method B. A mixture of 20 mmoles of the nitro substituted carboxylic acid and 10 mL of thionyl chloride were heated under reflux for three hours. Excess thionyl chloride was removed by distillation in vacuo (water aspirator). The resulting oil was placed under vacuum (1 torr) for 1h to remove the last of the thionyl chloride. The resulting yellow solid was dissolved in 20 mL of dry chloroform. A solution of 20 mmoles of N-hydroxysuccinimide in 20 mL of dry chloroform was added and the resulting solution was cooled in an ice bath. A solution of 1 mL of dry pyridine in 20 mL of dry chloroform was slowly added (1 h) and the resulting mixture was allowed to slowly warm to room temperature with stirring overnight. The chloroform was removed *in vacuo* and the resulting solid was sequentially tritiated with cold portions of 5% hydrochloric acid, water, and 5% sodium bicarbonate, then air dried.

Purification and characterization: The crude NHS esters were recrystallized to a constant melting from either chloroform-hexane or chloroform-acetone mixed solvents. Product purity was confirmed by proton nuclear magnetic resonance using a GE QE 300 spectrometer.

Preparation of Immunizing and Coating Antigen

Method C. A solution of 30 mg of the active NHS esters of 4-nitrophenylacetic acid or 2,4-dinitirophenylacetic acid in 6 mL of tetrahydrofuran was slowly added to a cooled(0°C), stirred solution of 60 mg of BSA or OVA dissolved in 6 mL of 0.10 M aqueous lithium borate buffer (pH 9). The deep red solution which formed was stirred overnight. The resulting pale yellow solution was dialyzed in a cold room at 2°C against several changes of 500 mL each deionized water. Any sediments were removed by centrifuging the solutions and the resulting protein solutions were transferred to small vials (1 mL) and stored in a freezer at -22°C. Approximations of hapten density on the protein were made by comparing the ultraviolet spectrum of the sample with the spectrum of starting protein and starting hapten.

Method D. The remaining haptens were conjugated by stirring a mixture of 5 mL of chloroform, 5 mL of 0.10 M lithium borate buffer (pH 9), 75 mg of BSA or OVA and 25 mg of the active NHS ester of the corresponding nitroaromatic hapten in a flask stored in a cold room (2°C) overnight. The solutions were centrifuged and the aqueous layers were then treated in the same manner as described in Method C.

Antisera

New Zealand white rabbits were injected subcutaneously at multiple sites with an emulsion consisting of 0.100 mg of BSA-hapten immunogen in 0.50 mL of phosphate buffer(PBS) and 0.50 mL of Freund's complete adjuvant. After 30 days a booster injection of 0.100 mg of immunogen in 0.50 mL of PBS buffer and 0.50 mL of Freund's incomplete adjuvant was given. Ten days after the boost the rabbits were bled and antibody titers were determined. In some cases additional boost were given after 10 day intervals to maintain high antibody titers.

General Enzyme-Linked Immunosorbent Assay Procedure

1. Microtiter plates(Immulon II) were coated by adding 100 µL of a solution 1 to 20 µg/mL of coating antigen in carbonate buffer (pH 9) to each well then storing the plates in a refrigerator (4°C) overnight.
2. The plates were emptied, washed with Saline-Tween solution, then tapped somewhat dry on paper towels.
3. Uncoated sites in the wells were blocked by adding 200 µL of a 0.5% ovalbumin in PBS-Tween solution and allowing the plates to stand 1 hr at room temperature.
4. The plates were emptied and washed with saline-Tween (3x).
5. A solution of antisera in PBS buffer was prepared by appropriate dilutions of previously frozen samples of rabbit antisera. After pre-incubation with standards, 100 µL samples were pipetted into the wells.

6. The plates were allowed to stand covered at room temperature for 2 hours then washed with saline-Tween (3x).
7. Biotinylated goat anti-rabbit conjugate (1mg/mL) was diluted by 1/2000 with PBS buffer, 100 µL/well was added to the plate, and the plate was incubated for 1 hr at room temperature.
8. The plate was emptied, washed with saline-Tween (3x), and partially dried.
9. A solution of avidin conjugated peroxidase (1mg/mL) was diluted by 6/1000 with PBS buffer; 100 µL was added to each well in the plate, and the plate was incubated for 30 min at room temperature.
10. A solution (100 µL/well) containing 1 mg/mL o-phenylenediamine and 1 µL/mL 30% hydrogen peroxide was added to the plate.
11. Analyses were done with a Molecular Devices microtiterplate reader interfaced to an IBM computer with Softmax software(Molecular Devices). Either the rate of color change (mOD/min) at 450 nm was recorded using the kinetic format or the final optical density (OD) was measured by difference of readings at 490 and 650 nm.
12. All readings were corrected for non-specific binding by using buffer as a blank.

Non-Specific Binding

Microtiter plates were coated by adding 100 µL of a solution containing 10 µg/mL of unconjugated proteins, BSA or OVA. The plates were incubated overnight then washed with saline-Tween. Several dilutions of antisera were prepared (1/250-1/20000) and added to the wells and steps 6-12 for the ELISA procedure were then performed. In all cases tested the OVA showed nearly blank background readings while, as expected, BSA showed strong binding with all antisera.

Checkerboard Titrations

The optimum dilution factors for the antiserum and coating antigen concentrations used in each assay were determined by checkerboard titration. Plates were coated by overnight incubation with coating antigen solutions (100 µL/well) ranging from 0.5 µg/mL to 20 µg/ml. Solutions of antisera (100 µL/well) with dilution factors ranging from 1/250 to 1/40000 were added to the wells and steps 6-12 of the ELISA were performed. A typical plot of sera dilution and coating antigen concentration is given in Figure 4 and antisera dilution values which will give a final OD reading of 1 at a coating antigen concentration of 1 µg/mL are given in Table I.

Competitive ELISA with Nitroaromatic Standards

For each analyte, a primary standard of 1.00 mg/mL of nitroaromatic compound in absolute ethanol was used to make a series of diluted standards in PBS-Tween buffer with concentrations ranging from 10,000 ng/mL to 0.1 ng/mL. An antisera solution of appropriate dilution (as determined from Table I) for the particular coating antigen to be used was prepared in PBS-Tween buffer. Samples of each nitroaromatic standard (1 part) were diluted 1/10 with antisera solution (9 parts)

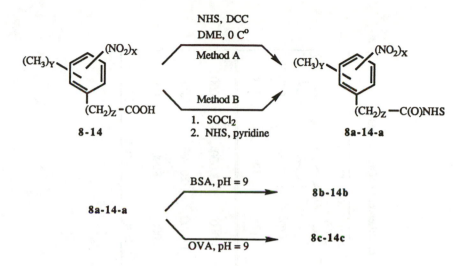

Figure 3. Synthesis of Immunizing and Coating Antigens.

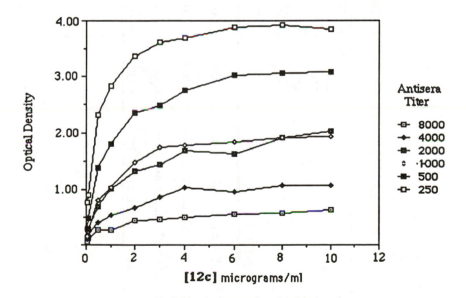

Figure 4. Typical checkerboard titration for antisera to **8b** from rabbit number 658 with coating antigen **12c** (System **D** in TableII).

Table I. A Comparison of Antisera Dilution Factors (Titers) for Different Coating Antigens

ANTISERA	Coating Antigens(compared at 1 µg/mL concentration)						
	8c	9c	10c	11c	12c	13c	14c
Rabbit 658/anti-**8b**	8,000	1,000	1,000	1,000	1,000	8,000	6000
Rabbit 659/anti-**9b**	12,000	16,000	16,000	500	12,000	20,000	1,000
Rabbit 1319/anti-**10b**	400	1,000	2,000	4000	8000	8000	4000
Rabbit 1320/anti-**11b**	750	>250	4000	8,000	500	6000	2000

test tubes and pre-incubated overnight at room temperature. One standard contained only antiserum and buffer (1/10) and was used as the control to determine the maximum kinetic or OD reading and another sample contained only buffer and was used as a blank. All steps of the general ELISA procedure were then performed. The relative rate of the ELISA with standards(V_{std}) was compared to the rate versus antisera alone(V_o) and values for the percent control (inhibition) were calculated by the formula V_o-V_{std}/V_o x 100. The percent control value was plotted against the log of nitroaromatic analyte concentration to construct typical standard curves (Figure 5-Figure 7).

Calculation of Comparative Inhibition

Cross reactivities for a variety of analytes with numerous combinations of antisera and coating antigens were tested. Initial screens were done using standards decreasing by ten from a concentration value of 10,000 ng/mL to a low value of 0.1 ng/mL. The results for several different analytes from the NPL list analyzed with the **11b**-antibody/**8c**-coating system are given in Figure 5. Similar experiments were performed using nitroaromatics analytes with several different immunogen and coating antigen combinations. The results of the ELISA analysis of trinitrotoluene with several different antisera-coating antigen systems are shown in Figure 6. In those cases where an analyte showed a promising inhibition range, additional standards were prepared and run in the range where inhibition curves displayed nearly linear behavior and the I_{50} (50% of the control) value was calculated by interpolation. Representative results for cross reactivity studies of several nitroaromatic analytes with seven ELISA systems are presented in Table II(**A-G**).

For those analytes that were successful competitors in the ELISA, multiple runs were performed by four different chemists over a six-month period; typical curves with error bars are shown in Figure 7. Typical standard deviation at each point was in the range of 10% of the control value at lower analyte concentrations and 2% at higher analyte concentrations. The standard deviations for the I_{50} values are shown in Table II.

Results and Discussion

Synthesis of hapten-protein antigens. The NHS esters in Figure 2 were successfully prepared by the reaction of the corresponding carboxylic acids with N-hydroxysuccinimide. Subsequent reaction with BSA and OVA led to the targeted hapten-protein conjugates. In each case, the ultraviolet spectrum of the conjugated protein showed that the number of lysines conjugated ranged from 30-50% of the total available lysines. Protein solutions could be stored at 0°C for long periods of time without any apparent deterioration.

Checkerboard titration of antisera and coating antigen. Figure 3 shows the two-dimensional determination of OD readings from ELISA runs in which antisera dilutions and concentrations of coating antigen were varied. At low coating and low antisera dilution, the final OD readings were low and the change in slope of the line insignificant. To optimize instruments readings, a coating antigen concentration and antisera titer were selected which gave an OD reading of near one after a 20 min reaction period for the peroxidase enzyme with hydrogen

Table II: Cross Reactivity Comparison for Various Analytes on Different Antisera/Coating Antigen Systems
(NI refers to non inhibition of ELISA system by that analyte)

SYSTEM	I50 VALUES FOR NITROAROMATIC ANALYTES (ng/mL)								
	1	2	3	4	5	6	7	14	15
A Rabbit 658/anti-**8b** **12c** coating antigen	NI	83±23	8000	NI		33	54±12	NI	NI
B Rabbit 658/anti-**8b** **10c** coating antigen	NI	25±17		NI		14	36±14	NI	
C Rabbit 658/anti-**8b** **11c** coating antigen	NI	550	NI	NI	NI	17	600	NI	350
D Rabbit 1320/anti-**11b** **8c** coating antigen	NI	5.9±3.0	33	NI	8.5	5.0±3.7	5.1±1.9	NI	NI
E Rabbit 659/anti-**9b** **8c** coating antigen	NI	NI		NI		NI	NI	NI	NI
F Rabbit 1319/anti-**10b** **8c** coating antigen	NI	1150	6000	NI	6000	NI	1200	9000	5500
G Rabbit 1319/anti-**10b** **9c** coating antigen	NI	6000	3500	NI	NI	NI	6500	NI	6000

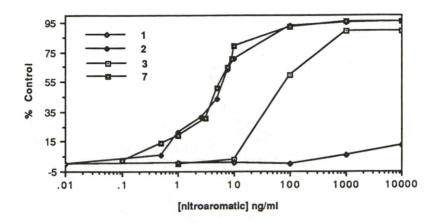

Figure 5.   Competitive ELISAs of four nitroaromatic analytes using antisera to **11b** from rabbit 1320 with **8c** as the coating antigen.

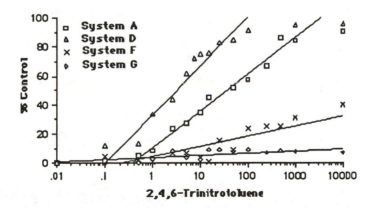

Figure 6.   Competitive ELISAs for four different combinations (Systems) of antisera and coating antigens given in Table II.

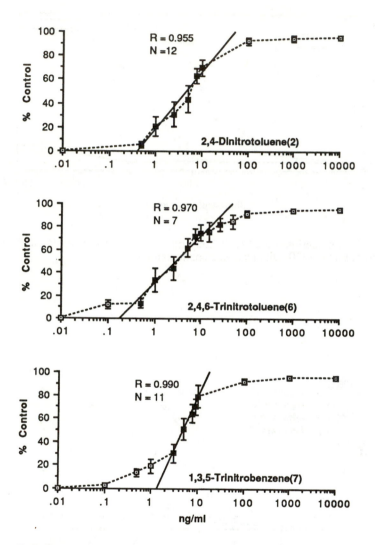

Figure 7. A Comparison of competitive ELISAs for three substrates, all with 1,3-dinitro substituents, with antisera to **11b** from rabbit 1320 with **8c** as the coating antigen (System **D** in TableII).

peroxide and o-phenylenediamine. For example, the data in Figure 3 suggests an antisera (**8b**) titer of 1/1000 with a coating antigen (**12c**) concentration of 1 μg/mL. Similar titrations were done with other combinations of antisera and coating antigens and Table I gives the antisera dilution required to give a final OD reading of 1 with a coating antigen concentration of 1 μg/mL. The data in Table I reflect binding affinities between antibodies and coating antigen and provides information regarding the structural features the antibody requires in binding to the hapten on the coating antigen. The homologous system in which the hapten on the immunogen and the hapten on the coating antigen were the same nitroaromatic(anti-**8b** on **8c**; anti-**9b** on **9c**; anti-**10b** on **10c** and anti-**11b** on **11c**) all had high titers, therefore strong binding, indicating a good fit between immunogen and hapten. Rabbit 659 (anti-**9b**) showed high affinity for nearly all of the nitroaromatic haptens, except **11c**, and is either specifically targeting the nitro functional group or is recognizing some other common part of the molecule such as the amide moiety linking the hapten to the protein.

Sensitivity of the ELISA to various analytes. Preliminary   screening of antisera/coating systems with various analytes employed standards with concentrations ranging from 0.1 ng/mL to 10000 ng/mL. Various systems were created using one of the coating antigen at a concentration of 1 μg/mL, and the antisera for a particular immunizing antigen obtained from one of the rabbits. The titer for the immunizing antigen was taken form the data in Table I. Standard ELISA procedures were used to measure antibody competition between the analyte and the hapten-protein conjugate used as coating antigen..

Either rate of color development (mOD/min) or final OD reading after twenty minutes was used to measure nitroaromatic antibody binding to the hapten coating antigen.   Therefore color intensity was inversely correlated to the concentration of nitroaromatic analyte. The presence of analyte capable of competitive binding and with the coating antigen for the antibodies reduced both the rate of color development and final OD reading. In order to standardize and compare the ELISA data, the rate (mOD/min) for each concentration of analyte was calculated as the fraction of rate for reading with antisera in the absence of analyte. These data are reported and plotted as a percent control and can be referred to as the percent inhibition in the ELISA. The value for 50% inhibition of the ELISA was used as a comparison point for different analytes on different immunogen/coating system. Figure 5 shows the results for ELISA analysis of nitrobenzene , 2,4-dinitrotoluene, 2,6-dinitrotoluene, and 1,3,5-trinitrobenzene with Rabbit 1320/anti-**11b** and coating antigen **8c** (System **D**,Table II). As can be seen from the figure, nitrobenzene did not act as an effective competitor (i.e., inhibitor) in the ELISA assay. Antibodies generated to the 3,5-dinitro-4-methyl phenyl moiety of immunogen **11b** clearly do not bind the nitrobenzene analyte at a level comparable to binding the 2,4-dinitrophenyl moiety of the hapten on the coating antigen. It is not surprising that the data show increased sensitivity when the system is used to test for 2,6-dinitrotoluene, since that analyte has identical structural features on the immunizing hapten. Interestingly, 2,4-dinitrotoluene and 1,3,5-trinitrobenzene are the analytes for which this system shows the greatest sensitivity. Apparently keeping the 1,3-dinitro arrangement intact in 2,4-dinitro toluene and 1,3,5-trinitrobenzene, but removing the steric effect of the methyl group found in 2,6-dinitrotoluene, has a beneficial effect on analyte/antibody binding.

Each analyte was also run with other coating/antisera combinations. Figure 6 shows the sensitivity of 2,4,6-trinitrotoluene to four different combinations of coating antigen and antisera. In each plot, only the linear region of the curve was used to draw a fit of the data. Antisera from rabbit 1320/anti-**11b** with coating antigen **8c** (System **D**) again shows excellent sensitivity for the analyte. Rabbit 658/anti-**8b** has a greater I_{50} (i.e., less sensitivity) but shows an excellent linear range from 1 ng/mL to nearly 1000 ng/mL. No combination of coating antigen with Rabbit 1319 /anti-**10b** allowed for effective competition by 1,3,5- trinitrotoluene.

Maximizing the sensitivity and precision   For those ELISA systems which showed a reasonable degree of sensitivity (I_{50} less than 100 ng/mL), efforts were made to maximize the sensitivity and investigate the precision of the results. For comparison purposes, changes in antisera dilution factors and coating antigen concentrations were made using the same set of nitroaromatic standards. Blocking by OVA after coating the plate was found to reduce noise and background in the readings. It was generally found that the scatter of the data as well as sensitivity (as measured by the I_{50}) improved when reading in the kinetic mode as long as values for the rate were near 100 mOD/min. Finally, extended incubation periods for the ELISA steps involving biotinylated anti-rabbit IGg and avidin peroxidase generally increased the rate and led to improved sensitivity.

Reproducibility.   Repeat   analysis were done for all ELISAs having a reasonable I_{50} (< 100 ng/mL). These experiments were run over a period of nine months, by different individuals and often with different preparations of coating antigen. Standard curves for 2,4-dinitrotoluene, 2,4,6-trinitrotoluene, and 1,3,5 trinitrobenzene analytes run with the same ELISA system, along with standard deviations data, are plotted in Figure 7. The agreement in data for each analyte was good. For example, the standard deviation in the data for 2,4-dinitrotoluene ranged from a high of 12% at a concentration of 2.5 ng/mL to a low of 1.3% at 100 ng/mL. A second feature of the data were the striking similarity in the curves for the three nitroaromatic analytes. It appears that a 1,3-arrangement of nitro groups is very important if the antibody is to successfully bind with the analyte. Composite curves were also used to calculate the mean I_{50} value as well as calculate its standard deviation. Similar data on other antisera/coating system were collected. Table II summarizes the final results from the systems studied.

Summary of results.   The data in Table II , in which seven different systems of antisera and coating antigen(**A-G**) were used for ELISA analysis of NPL nitroaromatics, reveals some interesting trends in the type of nitroaromatic compounds capable of being successfully detected by ELISA. Mononitroaromatics such as nitrobenzene, 4-nitrophenol, and 2-nitrotoluene did not inhibit any of the ELISA systems used in this study. In all cases, the I_{50} values were high (>9000 ng/mL) and the lower detection limits were too high to have practical analytical applications. This was even true on ELISA Systems **E, F,** and **G** which were originally designed to be "universal indicators" for nitroaromatics as all attempts to compete analyte against coating antigen with reasonable sensitivity were unsuccessful. The antibodies apparently recognized the single nitro group but perhaps the attachment to the immunizing protein was an important factor in tight binding between antibody and coating antigen. System **A-D** all showed a degree of successful competitive ELISA; however, system **D** (Rabbit 1320/anti- **11b** and coating antigen **8c**) was an order of magnitude better than any other system for detecting nitroaromatics with 1,3-dinitro functionality. It is clear that in this system the 1,3-dinitro functionality is an important feature needed for a successful competitive ELISA. This result was not unexpected in that the immunizing proteins also had that same structural feature.

Conclusions. The ELISA method may be a promising tool for analysis of several of the nitroaromatic compounds found at NPL hazardous waste sites. Detection of 2,4-dinitrotoluene, 1,3,5-trinitrobenzene, 2,4-dinitrophenol, and 2,4,6-trinitrotoluene at a lower limit of 1 ng/mL under the conditions of this ELISA suggest that direct detection of 10 ng/mL of these nitroaromatics may be accomplished in environmental samples without sample concentration. Investigations are underway to analyze soil and water samples to determine what, if any, matrix effects there will be in these ELISAs.

Acknowledgments

We gratefully acknowledge financial assistance from the Environmental Protection Agency.

Notice

Although the research described in this article has been supported by the U.S. Environmental Protection Agency, Environmental Monitoring Systems Laboratory, Las Vegas, NV (through assistance agreements CR-814709), it has not been subjected to Agency review and therefore does not necessarily reflect the views of the Agency and no official endorsement should be inferred.

Literature Cited

1. Harrison,R. O.; Gee, S.J. and Hammock, B.D. Immunochemical Methods of Pesticide Residue Analysis. In Biotechnology for Crop Production; Hedin, P. A.; Menn, J.J. and Hollingsworth, R. M.; Eds. ACS Symposium Series 379; American Chemical Society, Washington, D. C., 1988, pp 316.

2. Hammock, B. D. and Mumma, R. O. Potential of Immunochemical Technology for Pesticide Analysis. In Recent Advances in Pesticide Analytical Methodology; Harvey, J.,Jr. and Zweig, G.; Eds. ACS Symposium Series 136; American Chemical Society, Washington, D. C., 1980, pp 321-352.

3. Van Emon, J.M.; Seiber, J. N. and Hammock, B. D. Immunoassay Techniques for Pesticide Analysis. In Analytical Methods for Pesticides and Plant Growth Regulators, Vol XVII, 1989, pp 217-263.

4. Cheung, P. Y. K.; Gee, S. J. and Hammock, B. D. Pesticide Immunoassays as a Biotechnology. In The Impact of Chemistry on Biotechnology: Multidisciplinary Discussions; Phillips, M. P.; Shoemaker, S. P.; Middlekauf, R. D. and Ottenbrite, R. M.; Eds. ACS Symposium Series 362. American Chemical Society, Washington, D. C., 1988.

5. Jung, F.; Meyer, H. D. and Hamm, R. T. J. Agric. Food Chem., 1989, 37, 1183.

6. Hall, J. C.; Deschamps, R. J. A. and Kreig, K. K. J. Agric. Food Chem., 1989, 37, 981.

7.   Gee, S. J.; Miyamoto, T.; Goodrow, M. H.; Boster, D. and Hammock, B. D.
     J. Agric. Food Chem., 1988, 36, 863.

8.   Wie, S. I. and Hammock, B. D. J. Agric. Food Chem., 1984, 32, 1294.

9.   United States Environmental Protection Agency. National Priority List for
     Hazardous Waste Sites. October 1986.

10. Synthetic Organic Chemicals: United States Production and Sales. 1982
     USITC Publication # 1422. p 27, U. S. Government Printing Office,
     Washington, D. C. 1983.

11. Synthetic Chemicals: United State Production and Sales. 1985 USITC
     Publication # 1892. U.S. Government Printing Office, Washington, D. C.
     1986.

12. Rickert, D. E.; Butterworth, B. E. and Popp, J. A. CRC Crit. Rev. Toxicol.,
     1984, 13, 217.

13. Mori, M.; Miyahara, T.; Moto, K.; Fukukawa, M.; Kozuka, H.; Miyagoshi,
     M. and Nagayama, T. Chem. Pharm. Bull., 1984, 33, 4556.

14. Spanggord, R. J.; Mortelmans, K. E.; Griffin, A. F. and Simmon, V. F.
     Environ. Mutagen, 1982, 4, 163.

15. Dixit, R.; Schut-Herman, A. J.; Klaunig, J. E. and Stoner, G. D. Toxicol.
     Appl. Pharmacol., 1986, 82, 53.

16. Stryer, L. Biochemistry; W. H. Freeman: New York, 1988, 3rd ed; p 890.

RECEIVED June 5, 1990

Chapter 10

Avermectins

Detection with Monoclonal Antibodies

Alexander E. Karu[1], Douglas J. Schmidt[1], Carolyn E. Clarkson[1],
Jeffrey W. Jacobs[2], Todd A. Swanson[3], Marie L. Egger[3],
Robert E. Carlson[3], and Jeanette M. Van Emon[4]

[1]Department of Plant Pathology and [2]Department of Chemistry, University
of California, Berkeley, CA 94720
[3]ECOCHEM Research, Inc., Chaska, MN 55318
[4]Environmental Monitoring Systems Laboratory, U.S. Environmental
Protection Agency, Las Vegas, NV 89193–3478

Avermectins — macrocyclic lactone natural products with potent
antiparasitic, insecticidal, and antihelmintic properties — are
gaining increasing use in veterinary, medical, and agricultural
applications. Extraction and instrumental analysis methods for
avermectins are complex, lengthy, and expensive, making
immunoassay a desirable alternative. Using ivermectin 4''-
hemisuccinate-protein conjugates as immunogens, we prepared 20
monoclonal antibodies (MAbs) that detect avermectins at 3 to 400
ppb in competition enzyme immunoassay (EIA). The EIA tolerates up
to 20% (v/v) of several organic solvents, and detects avermectins A_2,
B_1, B_2, and B_1 monosaccharide and aglycone. The MAbs recognize
major substituents and structural differences associated with
biological activity. We are currently developing simplified methods
to extract avermectins for EIA of environmental samples and
commercial formulations.

The avermectin family of antibiotics, fermentation products produced by the soil
microorganism *Streptomyces avermitilis,* were discovered in the late 1970s.
Studies from several groups soon demonstrated that these compounds and the
related milbemycins were powerful cidal agents for numerous species of insects,
helminths, and parasites (1-2). The major mode of action involves interference with
muscular activity mediated by γ-aminobutyric acid — the major neurotransmitter
in these pests (3). One major form called abamectin has proven very effective as in
insecticide formulations (e.g., Avid), and it is now registered for use on citrus,
strawberries, cottonseed, and many other crops and ornamentals. Another form,
generically known as ivermectin, is gaining increasing use in formulations with
tradenames such as Ivomec and Eqvalan to treat livestock in the U.S. and
overseas (4). The USDA Food Safety Inspection Service has set an action level of
15 to 20 ppb for ivermectin and 50 to 75 ppb of total residues in beef and pork liver
(5). Ivermectin has perhaps become best known for its use in experimental human
studies sponsored by the World Health Organization to treat onchocerciasis, or

0097–6156/90/0442–0095$06.00/0

"river blindness," a filarial parasitic disease endemic in West Africa. The success of these studies has led some epidemiologists to predict that ivermectin therapy could eliminate the threat of river blindness in Africa within a decade (6). Although the avermectins are relatively non-toxic to higher animals and humans, the uncommonly wide range of applications raises concern about their impact on insect ecology and the development of resistant insects and parasites.

The avermectin family includes several analogs with different biological activities; most are roughly 80:20 mixtures of isomers which are prohibitively difficult to separate. The complete synthesis of avermectin A_{1a} has been reported (7). Unmetabolized ivermectin or abamectin, and their monosaccharides and aglycones, are the major residues in meat and agricultural commodities (1). These are generally analyzed by HPLC.

The parent compounds and the metabolites have very low aqueous solubility. They have proven difficult to extract from some matrices: the current method for ivermectin involves some 41 concentration and clean-up steps preceding HPLC (8). A Merck method for recovering abamectin residues from strawberries and formation of fluorescent derivatives for HPLC analysis has 18 separate steps (9). A recently published two-step solid-phase recovery procedure for ivermectin from serum indicates that it is possible to combine an abbreviated concentration and cleanup method with a sensitive and specific detection system - in this case, liquid chromatography (10). An immunoassay for avermectins that could be interfaced with simplified residue recovery protocols is a promising solution to the intensifying demands on regulatory agencies to monitor these compounds.

Methods

Details of the hapten and conjugate syntheses, the derivation of the hybridomas, and the optimization of the competition EIA will be presented elsewhere (Schmidt, D., Clarkson, C., Swanson, T., Egger, M., Carlson, R., Van Emon, J., and Karu, A., submitted to J. Ag. & Food Chem.).

Synthesis of Avermectin Haptens and Conjugates. Figure 1 shows the structure of the avermectin molecule. Its major features include a central macrolide ring, the spiro ring system where the R1 and R2 substituent groups are located, and the α-L-oleandrose disaccharide. Ivermectin (an approximately 80:20 mixture of 22,23-dihydroavermectins B_{1a} and B_{1b}) was used to synthesize haptens, based on variations of the chemistry described by Mrozik, et al. (1982) for the preparation of avermectin acyl derivatives. The site 1 hapten (4''-O-succinoylivermectin) was prepared in a three step protection-succinylation-deprotection sequence which utilized dimethylaminopyridine (DMAP) to effect the succinylation of the 4''-hydroxy group. The site 2 hapten (5-O-succinoylivermectin) was prepared by direct succinylation. These haptens were coupled to bovine serum albumin (BSA), conalbumin (CON), or keyhole limpet hemocyanin (KLH) using a modified 1-ethyl-3(3-dimethylaminopropyl)-carbodiimide protocol (11). The conjugates were purified by Sephadex gel filtration prior to determination of hapten density by spectrophotometry. The hapten densities ranged from 5 to 30 avermectin molecules per molecule of carrier. The conjugates were soluble at protein concentrations around 3 mg/ml in phosphate-buffered physiological saline. They remained stable under refrigeration for more than a year.

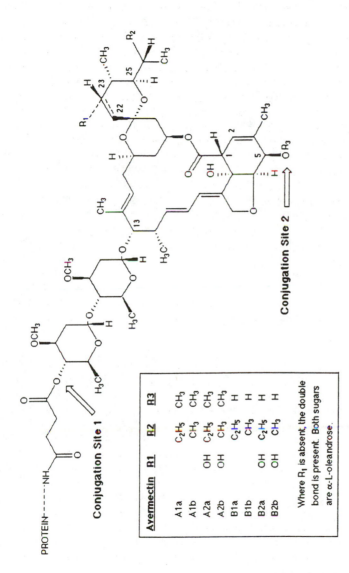

Conjugation Site 2

Conjugation Site 1

PROTEIN----NH

Avermectin	R1	R2	R3
A1a		C_2H_5	CH_3
A1b		CH_3	CH_3
A2a	OH	C_2H_5	CH_3
A2b	OH	CH_3	CH_3
B1a		C_2H_5	H
B1b		CH_3	H
B2a	OH	C_2H_5	H
B2b	OH	CH_3	H

Where R₁ is absent, the double bond is present. Both sugars are α-L-oleandrose.

Figure 1. Structure of ivermectin, showing the major analogs and the sites that were derivatized to form the hapten-protein conjugates used as immunizing and screening antigens. The 4" hemisuccinate ("Site 1") or the hemisuccinate formed through the oxygen on carbon 5 ("Site 2") were conjugated to carrier proteins as described in Methods.

Preparation of Hybridomas. We used several strategies to increase the chances of obtaining useful avermectin-specific antisera and MAbs. Pairs of Swiss Webster, Biozzi, and B10.Q mice were immunized with a total of 4 doses of one of the site 1 or site 2 BSA, CON or KLH conjugates over a six month period. The extended immunization schedule was necessitated by the very poor initial responses we observed to the site 1 conjugates, and negligible responses to the site 2 conjugates. We sampled sera from the mice at intervals of 3 to 4 weeks, and identified the best responders, based on the serum titer in direct EIA, and the lowest detectable dose (LDD), and half-maximal inhibition (I_{50}) for ivermectin in competition EIA. Mice immunized with the site 1 conjugates differed over tenfold in titer, LDD, and I_{50}, and we used only the best responders — two Swiss Webster mice and one B10.Q mouse — to prepare hybridomas. None of the mice immunized with site 2 conjugates responded well enough to be used for hybridoma preparation.

To further improve the chances of obtaining sensitive MAbs, we cloned the hybridomas as they were selected, by seeding the fused cells at 3.5×10^4 cells per well in 96-well plates. This cell density was low enough to give over 95% likelihood that only one hybridoma per well would develop, and not be overgrown by others of little or no interest. Of 9,888 wells (103 96-well plates) seeded, 1,686 colonies developed and were screened for reactivity toward ivermectin by EIA on plates coated with a conjugate different from the immunizing antigen. Of these, 485 recognized ivermectin, but only 36 proved to be stable antibody producers after several passages in culture. Competition EIA demonstrated that 33 of these were inhibitable by ivermectin, with I_{50} values from 3 ppb to over 400 ppb. The 5 hybridomas with the lowest I_{50} values were subcloned, and pools of the culture fluids were used for immunoassay development. All 5 were found to be of the $IgG_1\kappa$ immunoglobulin subclass.

Recovery and Analysis of Abamectin Residues.   Sample preparation was a modification of the initial steps of Merck & Co. Method 8001 (9). C_8 solid-phase extraction columns (Fisher Prep-Sep, 300 mg resin) were conditioned by consecutive washes with hexane, ethyl acetate, methanol, acetonitrile, and water (20 ml. each). Samples (100 ml of glass-distilled water) were spiked with various amounts of abamectin, adjusted to 25% (v/v) acetonitrile, and applied to the columns. The columns were washed with 10 ml of water, and then the abamectin was eluted in 12 ml of acetonitrile. The eluates were evaporated to 1.0 ml at 70° under nitrogen. Dilutions of these samples were made in PBS-Tween-20% acetonitrile, and mixed with equal volumes of MAb B11C2.1 (1:200 in the same buffer) in sealed 1.4 ml polypropylene tubes. After incubation for 2 hr to about 14 hr (overnight) at room temperature, replicate aliquots (100 µl) were transferred to Immulon 2 EIA wells coated with 25 ng ivermectin 4′′-hemisuccinate-CON for the standard competition EIA. Standard curves consisted of 8 dilutions of abamectin, from 0.01 ppb to 500 ppb, in triplicate.

Frozen strawberries were weighed and the frozen fruit was homogenized in a Waring Blendor at low speed. Aliquots of 10 gm. of homogenate were spiked with various amounts of a 10 ppm stock of abamectin in acetonitrile. One gram of celite filter aid and 25 ml of acetonitrile:H_2O :: 4:1 was added to each aliquot, and the mixtures were homogenized for 30 sec at medium speed in a Polytron homogenizer.

The homogenates were filtered through a double layer of Whatman #4 paper wetted with 2 ml of acetonitrile on a Buchner funnel. The filtrates were collected under slight vacuum in silanized glass 250 ml filter flasks. The Polytron vessel was rinsed with 25 ml of acetonitrile : H2O :: 1:4, followed by two rinses with 2 ml of acetonitrile. Each of these rinses in turn, and a final 3 rinses with 2 ml of acetonitrile were added to the filter cake, all of the filtrates were combined, and 50 ml of water was added to reduce the acetonitrile concentration to roughly 30% v/v. C_8 columns were conditioned, the samples were applied, and abamectin was eluted and concentrated as described above for spiked water samples.

Results

Enzyme Immunoassay of Avermectins.

The MAbs proved to be usable in a standard competition EIA, in which microplate wells were coated with a small amount of ivermectin hemisuccinate-protein conjugate, and the analyte competed with this immobilized ivermectin for binding with a limiting amount of the MAb. Bound MAb was detected by incubating the plates first with an enzyme conjugated to anti-mouse antibody, then with a chromogenic substrate for the enzyme. The sigmoidal dose-response curves, inversely related to the amount of avermectin analyte, were fitted with a 4-parameter logistic equation, and unknowns were interpolated from the standard curves. The procedure is illustrated in Figure 2.

For maximum precision and reproducibility, we found that it was necessary to incubate the standards and unknowns with the MAb for at least 2 hr at room temperature before the mixtures were added to the coated EIA wells. The dose-response curve of the EIA was between 1 and 100 ppb for ivermectin, abamectin, and the other analogs and metabolites described below. We did not determine the ultimate detection limits of the procedures for abamectin residues in water or strawberry homogenate. However, we recovered as little as 0.1 ppb of abamectin in water with ≥90% efficiency by the procedure described. The assay is extremely economical; each sample well is coated with ivermectin-carrier conjugate containing about 25 ng of carrier protein, and each sample requires only 0.5 to 1 μl of MAb in filtered hybridoma culture medium.

Since extractions with organic solvents are essential to recover avermectin residues and keep them soluble, we investigated the ability of the MAbs to bind ivermectin and abamectin in the standard EIA buffer supplemented with various amounts of solvents commonly used to extract avermectins. All 5 MAbs reacted equally well with ivermectin in phosphate-buffered physiological saline containing 0.05% Tween 20 surfactant ("PBS-Tween") with or without 5% (v/v) methanol. C4D6, the MAb which we routinely used to detect ivermectin, worked equally well in PBS-Tween containing 5% or 10% dimethyl sulfoxide, 5% tetrahydrofuran, as much as 20% dimethyl formamide or acetonitrile-25 mM H_3PO_4, or up to 30% acetonitrile or methanol. This tolerance of organic solvents should simplify use of the EIA with liquid- and solid-phase extraction methods for residue analysis.

BSA and CON conjugates of ivermectin were not affected by freezing, and EIA plates coated with these conjugates could be stored frozen. The mean I_{50} for ivermectin varied from 1 to 2.3 ppb in 5 EIAs conducted on plates that were coated and stored frozen for periods up to 30 days (Figure 3). EIA results were very similar with 7 brands of commercially available EIA plates, and with commercial

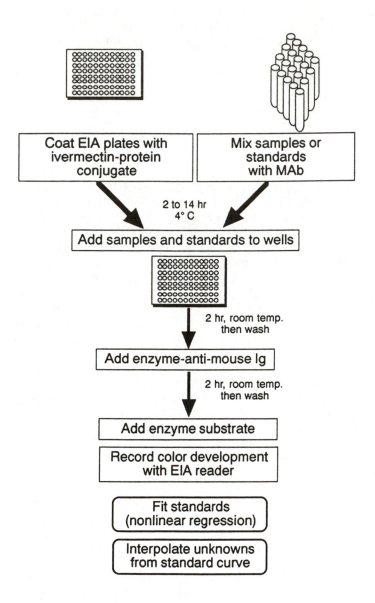

Figure 2. Flow chart for competition EIA of avermectins. All immunoreagents and other materials, except the ivermectin-protein conjugates and MAbs described in this paper, are commercially avaliable.

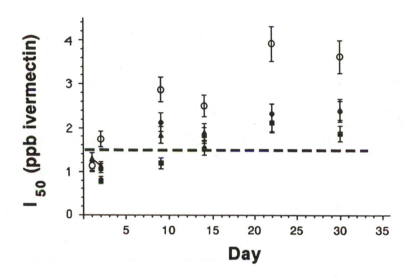

Figure 3. Reproducibility of the ivermectin competition EIA using plates coated the night before each assay, using conjugates stored at 4°C (open symbol), or plates coated the day before the study began (Day 0) and stored at -20°C until the day they were used (closed symbols). All plates were coated with ivermectin 4″-hemisuccinate-CON (25 ng carrier/well), and the MAb was C4D6. (o) Plates coated the night before they were used, with conjugate kept at 4°C. (▲), Plates coated, then emptied, and stored dry at -20° until the day of assay. (■), Plates coated and stored at -20° with the coating antigen in coating buffer left on; (●), plates coated, emptied, the wells refilled with coating buffer, and stored at -20°; Each data point is the mean ± standard error for I_{50} values fitted from 3 EIA plates, with 8 replicate curves run on each plate on the day indicated.

"detecting conjugates" of goat anti-mouse immunoglobulin coupled to alkaline phosphatase, urease, or horseradish peroxidase.

Specificities of the Monoclonal Antibodies. We examined the specificities of the MAbs in a series of competition EIAs that measured the ability of various avermectin analogs to compete with ivermectin site 1 or site 2 conjugates immobilized on the plates. Table I summarizes the I_{50} values obtained with 5 avermectin analogs, abamectin monosaccharide, and ivermectin monosaccharide and aglycone. Although small differences (e.g., the values obtained with abamectin obtained from two different sources) did not appear to be significant, distinct patterns were evident in the larger differences in reactivity of each of the 5 MAbs toward the different analogs. For example, MAb B11C2 reacted least well with avermectins B_2 and A_2, which were mainly distinguishable from the others because they have an OH group at R1. The difference is not just a simple —H, —OH substitution; it involves sp^2 to sp^3 hydration of the 22, 23 double bond, significantly altering the van der Waals shell in this part of the molecule. B11C2 also bound better to abamectin (avermectin B1) than to avermectin B2, while the other 4 MAbs behaved oppositely, binding as well or better to avermectin B2 than they did to avermectin B1.

MAbs B2A2 and C1A3 bound well to all of the analogs except avermectin A_2, which was not recognized by either of these antibodies. Furthermore, on plates coated with ivermectin conjugated through site 2, the rates of the EIAs with B2A2 and C1A3 at the LDD were about one tenth the rates observed on plates coated with ivermectin conjugated through site 1 (data not shown). This indicated that a CH_3 group or tether at R3 blocked recognition by these antibodies. MAbs C4D6 and C5D6 bound much better to ivermectin and its monosaccharide than to abamectin and its monosaccharide, although the only difference seen in the conventional 2-dimensional representation of these structures, was the olefinic 22,23 double bond in abamectin. Finally, we observed that the I_{50} values for all of the antibodies and nearly every analog were smaller, i.e., the EIA was more sensitive, on plates coated with ivermectin conjugated through site 2 than on plates coated with the site 1 conjugates. This suggested that all of the MAbs bound more tightly to the site 1 conjugates.

To gain a better idea of how the specificity EIA data related to avermectin structure, we modeled the idealized gas-phase structures of the analogs with MacroModel 2.0 software on a MicroVAX linked to an Evans & Sutherland Picture System 300 (12). After verifying the chemical structures visually, we derived 3-dimensional representations of the minimum free-energy conformations using an iterative MM2 algorithm, which took into account bond stretching, angle bending, torsion, Van der Waals, electrostatic, and non-bonding interactions. The electrostatic interactions were calculated using partial atomic charges, rather than the dipole-dipole interactions used by the standard version of MM2 (13). Each structure was refined to a first derivative root-mean-square free energy less than 0.5 KJ/Å.

These 3-dimensional structures are shown in Figure 4. The top panel shows avermectin B_{1a} and 22, 23 dihydroavermectin B_{1a}, from which we can see some general features of abamectin and ivermectin that are not obvious from the 2-dimensional representation. These include the rigid, "corrugated planar" form of

Table I. Specificities of Monoclonal Antibodies for Various Avermectins. The values are the amount of the indicated compound giving half-maximal inhibition in a competition EIA on plates coated with ivermectin 4''-hemisuccinate-conalbumin (coating antigen 1) or ivermectin-C_5-hemisuccinate-BSA (coating antigen 2) at 25 ng of carrier protein per well. All analogs were approximately 80:20 mixtures of the respective isomers, e.g., Avermectin A_2 was a mixture of avermectin A_{2a} : avermectin A_{2b} :: 80:20. Stock solutions of all compounds in methanol were standardized by spectrophotometry (see refs. 14-17 for extinction coefficients and chemical data), and dilutions for assay were made in PBS-Tween-5% methanol. Results are the mean of 8 replicates, and standard errors were approximately ±5% of the means. NR = not recognized in amounts up to 10 ppm

MAb --->	B11C2		B2A2		C1A3		C4D6		C5D6	
Coating antigen--->	1	2	1	2	1	2	1	2	1	2
Analyte in liquid phase	<-----				I_{50}	(parts per billion)			---------->	
Ivermectin (22,23dihydroavermectin B1)	14.9	11.4	8.0	4.3	31.4	11.1	13.5	7.8	19.9	13.5
Ivermectin monosaccharide	9.3	6.4	11.3	7.2	20.6	9.4	7.4	7.2	28.9	14.7
Ivermectin aglycone	18.4	18.0	8.1	7.8	31.2	15.2	53.9	57.9	82.6	53.9
Abamectin (avermectin B1) — source #1	13.4	10.2	6.7	5.0	12.7	8.4	57.9	36.2	71.9	54.8
Abamectin (avermectin B1) — source #2	21.0	10.2	9.4	5.1	16.8	6.9	74.4	45.2	79.3	44.1
Abamectin monosaccharide	9.6	8.2	7.0	5.7	12.1	7.7	44.1	29.5	71.5	34.7
Avermectin B2	61.9	47.3	3.7	2.6	6.8	4.0	10.9	6.0	30.7	12.3
Avermectin A2	54.5	66.3	NR	160.7	NR	NR	25.8	15.4	51.2	24.3
4''deoxy-4''epi methylamino-avermectin B1 hydrochloride	5.7	3.6	3.0	2.0	4.4	3.4	23.1	12.7	54.9	15.6

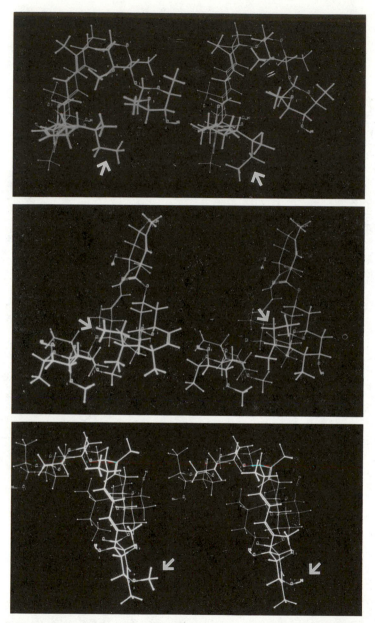

Figure 4. Molecular models of avermectins used in the specificity studies of Table I. Top panel; avermectin B_{1a} (left) and 22,23,dihydroavermectin B_{1a} (right). Middle panel; avermectins B_{1a} (left) and B_{2a} (right). Note differences in the $C_{22} - C_{23}$ bonding, and its effect on orientation of the R2 ethyl group (arrows) in these free energy-minimized structures. Bottom panel; avermectins A_{2a} (left) and B_{2a} (right), oriented to show the OCH_3 or OH groups at R3 (arrows).

the macrolide ring, and the "cupped" or "inverted L" shape due to the bent disaccharide moiety. Over-all, the conformations are nearly identical; virtually the only differences are the 22, 23 bond on the spiro ring, and the change it causes in orientation of the R2 ethyl group. These differences were clearly recognized by MAbs C4D6 and C5D6.

The middle panel of Figure 4 compares avermectins B_{1a} and B_{2a}. Here again there is near conformational identity, except for the olefinic 22,23 double bond in avermectin B_{1a}. The spiro ring in avermectin B_{2a} resembles that of the immunizing antigen, ivermectin, and the orientation of the R2 ethyl group is again different, depending on the ring bonding. MAb B11C2 was apparently able to recognize this difference. The bottom panel shows avermectins A_{2a} and B_{2a}, which appear to have no conformational differences other than substitution of the o-methyl for the hydroxyl group at R3 that abolished the binding of MAbs B2A2 and C1A3.

Recovery and analysis of avermectin residues. The development of specific protocols for avermectin residue analysis by EIA is under way, but at an early stage. Work at EMSL-Las Vegas has emphasized development of a greatly simplified procedure for recovering ivermectin residues from liver. At Berkeley, we have focused on an abbreviated extraction of abamectin residues from strawberries. Schematic diagrams of the procedures we are currently evaluating are shown in Figures 5 and 6. In both cases, the major concerns are the efficiency of the extraction procedures and the types of matrix effects encountered. Recoveries of ivermectin from liver have to be improved, but the procedure of Figure 5 appeared to be relatively free of matrix effects (data not shown). Tables II and III summarize the recoveries of abamectin from spiked water and strawberry homogenates using the protocol described in Methods and summarized in Figure 6. Recoveries in this procedure appeared to be acceptable for routine use, but there is matrix interference in the strawberry extract, manifested as a "column blank" of 3 to 5 ppb in the absence of a spike. We are currently attempting to improve the liver and strawberry residue extraction methods and bring them to the application stage.

Discussion

The MAbs and EIA we have developed provide a means of detecting avermectins at the level of sensitivity now possible with HPLC and gas chromatography. The EIA employs standard methods, and it is stable, reproducible, and economical. The solvent tolerance exhibited by the MAbs makes the assay compatible with methods for recovering residues of these very hydrophobic compounds. Monoclonal antibodies have the well-known advantages of defined affinity, specificity, and potentially "infinite" supply. The IgG_1 subclass of all of the avermectin MAbs makes them easy to purify to near-homogeneity by well-documented methods. This could facilitate tests of their usefulness in other immunoassay formats.

It is a common misconception that immunoassays with polyclonal sera are likely to be more sensitive than MAbs for immunoassay of molecules the size of the avermectins. The I_{50} values of the sera from the best-responding mice in our study were between 20 and 50 ppb, while the 8 most sensitive MAbs had I_{50} values of 3

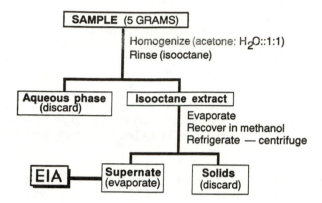

Figure 5. An abbreviated method for recovery of ivermectin residues from liver.

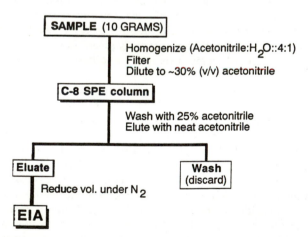

Figure 6. Flowchart for extraction of abamectin residues from strawberries.

Table II. Recovery of abamectin from spiked water samples taken through the extraction protocol of Figure 6. Values for the samples are the mean and standard error for triplicates that fell within the working range of the standard curve (0.2 to 0.7 of the normalized response), which had an I_{50} of 8.1 ppb. Values preceded by the ≤ sign were below the limit of detection, i.e., they had a normalized response above the standard curve limit of 0.7.   The "column blank" is the mean ± SE of the unspiked samples

Spike-->	0 ppb	0.96 ppb	1.92 ppb	4.8 ppb
Replicate				
1	0.1 ± 0.04	0.9 ± 0.03	1.8 ± 0.01	5.1 ± 0.22
2	≤ 0.02	0.8 ± 0.09	1.9 ± 0.03	4.8 ± 0.17
3	≤ 0.06	0.8 ± 0.03	1.7 ± 0.06	4.9 ± 0.17
ppb recovered (mean ± SE)	0.1±0.04	0.8 ± 0.05	1.8 ± 0.03	4.9 ± 0.19
— ["column blank"]		0.7 ± 0.05	1.7 ± 0.04	4.8 ± 0.19
% recovery		72.9%	88.5%	100.0%

Table III. Recovery of abamectin spikes from strawberry homogenates. Residue-free strawberries (California Certified Organic Farmers, Inc.) were purchased locally. Homogenates with the indicated spikes were processed through the extraction protocol of Figure 6. Values for the samples are the mean and standard error for triplicates that fell within the working range of the standard curve (0.2 to 0.7 of the normalized response), which had an I_{50} of 13 ppb. Values preceded by the ≤ sign were below the limit of detection, i.e., they had a normalized response above the standard curve limit of 0.7. The "column blank" is the mean ± SE of the unspiked samples

Spike-->	0 ng/g	4.8 ng/g	9.6 ng/g	19.2 ng/g	47.8 ng/g
Replicate					
1	4.3	8 ± 0.7	17 ± 0.7	24 ± 0.9	55 ± 0.8
2	5 ± 0.7	8 ± 0.7	12 ± 0.6	24 ± 1.3	46
3	3.3	9 ± 0.5	14 ± 1.5	20 ± 0.3	43 ± 4.1
ppb recovered (mean ± SE)	4 ± 0.7	8 ± 0.6	14 ± 0.9	23 ± 0.8	48 ± 2.9
− ["column blank"]		4 ± 0.6	10 ± 0.9	19 ± 0.8	44 ± 2.9
% recovery		83.3%	104.2%	99.0%	92.1%

to 12 ppb for ivermectin and abamectin. We speculate that EIAs using whole antisera to avermectins may not be as reproducible as assays using MAbs. This is because avermectin-protein conjugates elicit antibodies with a variety specificities, including some that bind strongly to the hapten but show little or no recognition of free avermectins.

It has been long known and well documented that small haptens give rise to MAbs with diverse affinities and specificities. Nevertheless, we were intrigued to find that the 5 most sensitive MAbs each had different specificities for the avermectins. It would be interesting to ask whether the specificities of the MAbs are related to biological activities of the avermectin analogs, but we have not attempted to do this.

The 3-dimensional molecular models of the avermectins helped us to define the subtle structural differences recognized by the MAbs, but these models must be viewed and interpreted with caution. Since we could not invoke solvent interactions, we have no way to know what the differences may be between these gas-phase representations and the conformations assumed in solution. On the other hand, data for other hydrophobic compounds, including some macrolide antibiotics, suggests that their gas-phase and solution structures are very similar, and the major differences we observed in Figure 4 were in hydrophobic parts of the molecule. For this reason, we did not attempt to model the last analog in Table I, the 4″ deoxy-4″epimethylamino avermectin B_1 hydrochloride, even though it was the best competitor in the EIAs, because it is far more water soluble than the others. Another concern is that the samples compared in the competition EIAs were mixtures of isomers, e.g., abamectin is actually avermectin $B_{1a}:B_{1b} :: 80:20$. Because it is virtually impossible to obtain the purified individual isomers for the EIA or other tests of antibody specificity, we can only assume that the competition EIA results primarily reflected how each MAb bound to the major isomer.

The competition EIA results, considered in the context of the molecular models, suggest that 3 domains on the avermectins influence recognition by the different MAbs. One domain, which includes C_5 and R3, profoundly affected binding by B2A2 and C1A3. The 22,23 bond in the spiro ring system and the oleandrose sugar proximal to the macrolide ring strongly affected binding by C4D6 and C5D6, while the spiro ring bonding and R1 most changed the reactivity of B11C2. The specificity results make predictions that could be used to test our interpretations, if additional analogs were available for the experiments. For example, MAbs B2A2 and C1A3 should bind poorly if at all to avermectin A_1, which, like avermectin A_2 has a methyl group at R3. The MAbs that recognize ivermectin aglycone might be able to recognize the milbemycins, highly similar structures which also have insecticidal activities. In any case, other immunochemical and physico-chemical techniques would be needed to rigorously define the way that these MAbs bind avermectins.

Our initial results indicate that greatly simplified and shortened extraction schemes are feasible for EIA of ivermectin in meat and abamectin in strawberries. However, considerable work remains before the liver and strawberry residue recovery methods are reduced to standard protocols, tested, and validated. A temporary tolerance level of 100 ppb for abamectin in citrus was set by the EPA in July 1988 (40 CFR 1, sec. 186.300). The Section 18 Emergency Exemption for use of abamectin on strawberries in California had a tolerance of 20 ppb. The

California Dept. of Food and Agriculture's HPLC detection limit for abamectin residues is 1 ppb, and contaminated strawberries are likely to have on the order of 2 to 5 ppb of residue. (Dr. Mark Lee, Calif. Dept. of Food & Agriculture, personal communication to Dr. Karu). The first two spikes in Table III are thus comparable to residues that could be left on field samples. Given the efficiency of recovery of the spikes and the errors, it is reasonable to expect that the EIA method could detect 2 to 5 ppb above the "column blank."

A simple first strategy to recover residues for EIA is to try existing multi-residue methods, or to use solvents and solid-phase extraction media that have proven satisfactory in the first steps of more involved instrumental analyses. This approach makes the transition from instrumental methods to EIA easier for technical personnel, and it facilitates confirmatory instrumental analysis during method validation. The uncommon structure of avermectin antibiotics makes it very unlikely that the EIA would detect compounds other than an avermectin in multi-residue extracts, though this remains to be tested. The greater concerns are completeness of the extractions, efficiency and reproducibility of recoveries, and freedom from non-specific matrix effects and interference by solvents.

The avermectins have been known for only about a decade, but this family of antibiotics has proven to be virtually non-toxic to humans and enormously beneficial in a surprisingly wide variety of medical, veterinary, and agricultural applications. As their use increases, so will their dispersal in the environment, and the exposure of consumers and agricultural workers. We are hopeful that the MAbs and EIA we have developed, in conjunction with simple, efficient residue recovery methods, will fill the growing need for rapid, inexpensive monitoring of avermectins in large numbers of samples.

Acknowledgments

We thank Merck, Sharpe, & Dohme for providing us with avermectin analogs and reference standards, Professor Peter Schultz for generously making his molecular modeling facilities available to us, and Dr. John G. MacConnell of Merck & Co. for valuable information and discussions on method development. J. Grassman and M. Bigelow provided excellent technical assistance in various stages of this work. This research was supported by EPA Cooperative Agreement CR-814619 and NSF Grant DIR 88-43099 (A.E.K.).

Literature Cited

1. Fisher, M.H.; Mrozik, H.; In Macrolide Antibiotics; Academic Press: New York, 1984; Chapter 14
2. Mrozik, H.; Eskola, P.; Fisher, M.H.; Egerton, J.R.; Cifelli, S.; Ostlind, D.; J. Med. Chem. 1982, 25, 658-663
3. Wang, C.C.; Pong, S.S.; Prog. Clin. Biol. Res. 1981; 97, 373
4. Benz, G.W. The Southwestern Entomologist 1985, Suppl. 7, 43-49
5. USDA Food Safety and Inspection Service; Compound Evaluation and Analytical Capability National Residue Program Plan. Brown, J., Ed.; USDA FSIS Science Program, Wash. D.C., 1987. p. 4.13

6. Eckholm, E.; New York Times Magazine Jan. 1989
7. Danishefsky, S.J.; Armistead, D.M.; Wincott, F.E.; Selnick, H.G.; Hungate, R. J. Am. Chem. Soc. 1989, 111, 2967-2980.
8. USDA Food Safety and Inspection Service: Chemistry Laboratory Guidebook, Sec. 5.035
9. Cobin, J.; Wehner, T.; Czeh, K.; Macaoay, A.; Rosenthal, H.; Tway, P.; 1988; Method 8001, Merck, Sharpe & Dohme, Three Bridges, NJ.
10. Oehler, D.D.; Miller, J.A. J. Assoc. Off. Anal. Chem. 1989, 72, 59
11. Weetall, H.H. In Methods in Enzymology: Ed. Mosbach, K.; Academic Press: New York, 1976; vol. XLIV, p. 139-148
12. Still, W.C.; Mohamadi, F.; Richards, N.G.J.; Guida, W.L.; Lipton, M.; Liskamp, R., Chang, G.; Hendrickson, T.; De Gunst, F.; Hasel, W.; MacroModel v.2.5, Dept. of Chemistry, Columbia University; New York, N.Y., 10027.
13. Allinger, N.L. J. Am. Chem. Soc. 1977; 99, 8127
14. Chabala, J.C., et al.; J. Med. Chem. 1980; 23, 1134-1136
15. Merck Index; 1985; 10th ed.; p. 753
16. Merck Index; 1985; 10th ed.; p. 128
17. Mrozik, H., et al.; J. Org. Chem. 1982, 47, 489-492

RECEIVED August 3, 1990

Chapter 11

Immunochemical Technology in Environmental Analysis

Addressing Critical Problems

Bruce D. Hammock, Shirley J. Gee, Robert O. Harrison, Freia Jung,
Marvin H. Goodrow, Qing Xiao Li, Anne D. Lucas, András Székács,
and K. M. S. Sundaram

Department of Entomology and Department of Environmental Toxicology,
University of California, Davis, CA 95616

Immunochemical technology is at a critical stage in its
development for use in environmental analysis. Primary
problems and issues regarding assay development and
applications such as outlining common misconceptions,
choice of format and choice of monoclonal or polyclonal
antibodies are discussed. More far reaching concerns
for the acceptance of the technology such as the roles
of government and industry in assay development and
standardization are also discussed. A committee to
coordinate the development of immunoassay in the
environmental field is proposed and its functions
outlined. How some of these problems are currently
being addressed are illustrated by work presented at
this symposium. Work from our laboratory illustrates
our approach to dealing with real world samples, more
difficult target compounds and complex matrices and
applying immunoassays to samples other than
environmental samples.

The next few years will be critical in the development of
immunochemical technology for use in environmental analysis. In
this light this manuscript has three objectives. The first is to
address how the critical problems facing the technology can be
approached. The second is to introduce aspects of this symposium by
pointing out how various laboratories are approaching these
problems. Finally, this manuscript will review briefly some of the
topics being addressed by this laboratory.

Evolution of Problems Facing Immunoassay

Changes in the Last Ten Years. Based on both the Miami American
Chemical Society (ACS) meeting 11 years ago and the ACS meeting 10
years ago, this symposium certainly has historical significance to

our laboratory. At the Miami ACS meeting in 1978 only two papers were presented on the immunoassay of pesticides. One was on our immunoassay for the optical isomers of the pyrethroid S-bioallethrin and the other was an immunoassay of parathion. The later study was from the laboratory of the late C. D. Ercegovich who was one of the early leaders in this field. The next year at the 1979 ACS meeting in Washington D. C. Drs. Harvey and Zweig requested that the laboratory address the potential of immunoassay for pesticide residue analysis in a symposium on *Recent Advances in Pesticide Analytical Methodology* (1). As one might anticipate, this talk drew a great deal of criticism based on many misconceptions regarding immunoassay. Some things certainly have changed in the ten years leading to this 1989 ACS meeting. Simply the increase from one paper to 27 papers on the immunochemical analysis of pesticides and other environmental chemicals illustrates a major change in the interest of pesticide chemists in immunoassay. This change has not been due to some magical improvement in immunochemical technology. In fact, immunochemical technology as it applies to the analysis of small molecules in environmental samples has not changed greatly in the last 10 years, while great changes have been made in chromatographic and spectral detection systems. The change has been in an increased awareness of the capabilities of immunoassays in the environmental field.

Misconceptions, Then vs Now.   At first glance this increase in interest in immunoassay might indicate that pesticide immunoassay has matured and left its doubters and problems behind. Such is not the case. Although there is wide interest in the technology and most agricultural chemical companies have in-house expertise in immunochemical technology, immunoassays have not been used to register a single pesticide nor is an Association of Official Analytical Chemistry (AOAC) validation of immunoassays for pesticides a common phenomenon. In fact, the technology seems to have traded one set of problems for another. Acceptance of the technology was stifled for years because many scientists concluded that immunochemistry had no place in environmental chemistry based on little appreciation of its power. We now find the major problem facing the technology is that it is being over sold, in some cases as a panacea, by people who do not understand the limitations of the technology.

A major theme of this book could be the same one we advocated ten years ago. That is that immunochemistry represents a very powerful analytical tool which is applicable to many but certainly not all problems in environmental chemistry. Thus, it complements but does not replace other analytical methods. The technology is so powerful and versatile that it should be in the repertoire of every analytical chemist. Yet there must be an understanding that the technology is very useful for some compounds and some problems, but that it is no panacea. Ten years ago the challenge was to encourage understanding of the tremendous power of the technology, while today we must preach the limitations.

The major problem facing the technology ten years ago was a lack of understanding of immunochemical analysis by environmental chemists. At the time the misconceptions seemed challenging to overcome, but in many ways they are less serious than the problems that the technology faces today. One significant problem was the jargon that surrounded immunochemical technology. A major role of the scientists involved in advocating immunoassay in the environmental field over the last decade has not been to pioneer new immunochemical technology but rather to translate the jargon used in clinical immunochemistry to the jargon used in environmental chemistry. As pointed out by Ken Hunter formerly of Westinghouse Bioanalytical, our job became dramatically easier with the advent of the microcomputer which generated a standard curve for analyte vs response. Such standard curves appear more familiar to analytical chemists, reassuring them that they are dealing with real analytical chemistry and not some qualitative biological phenomenon.

Another of the major misconceptions ten years ago was that immunoassays were bioassays, and some even thought that a rabbit died each time an analysis was run. We had a generation of analytical chemists who had fought to have chromatographic methods accepted over bioassays for residue analysis, and immunochemistry seemed to offer a great leap backwards. The realization that immunoassays are physical assays which simply use biological reagents is now wide spread. The fact that the cyclodiene antibodies published by Langone and Van Vunakis in 1975 (2) are still in use, is an excellent illustration that if properly handled immunochemical reagents are very stable.

A new misconception now exists in some quarters in this decade of biotechnology. Ten years ago the biological source of antibodies tainted immunoassays as a poorly reproducible black art practiced by biologists and not by real chemists. Now in some quarters the biological source of antibodies seems to impart magical qualities to immunoassays. Some people indicate that these assays can detect biological effects, but like any physical assay, immunoassays detect molecules which may or may not be associated with biological activity. Certainly the specificity of an immunoassay can correlate with that of a receptor molecule. However, such correlations are incidental. There is an effort to apply immunoassays to all compounds and problems with no appreciation for the technology's limitations or the strengths of competing technologies.

A major challenge facing all competent analytical chemists is to make sure that the technology is advocated based on its real strengths. If the technology is over sold based either on ignorance or on a desire to advance a product for profit or one's career, there is certain to be a backlash when immunochemistry fails to provide magical results.

There also is the indication that immunoassays allow untrained analysts to run highly sensitive assays. Although immunoassays may be very forgiving and easy to perform, the quality of the data generated for any physical assay will depend upon the

integrity of the samples and the skill of the analyst. As the
assays used become more difficult and the limit of detection lower,
the skill of the analyst must be greater as with any analytical
system.
        In some ways a more difficult problem has been that the
reluctance of residue chemists to embrace immunoassay has led to
the development of immunoassays for environmental chemicals in
metabolism, biotechnology, or clinical laboratories. Among this
group of scientists there sometimes is a reverse arrogance towards
the residue chemist who failed to adopt this technology. However,
the assays developed outside of an analytical laboratory often use
such simplistic hapten design that key recognition sites are
masked. Also, there is a vast difference in matrix effects between
clinical and environmental samples. The use of enzyme linked
immunosorbant assays (ELISA's) or other immunoassays in a real
analytical program will normally reveal new matrix problems totally
unfamiliar to the clinical chemist. The experience in this
laboratory is that a good analytical chemist can be trained to
perform ELISA in a matter of days. However, the conversion of an
immunochemist into an environmental chemist represents a major
change in career and philosophy. The collection, handling, and
processing of samples as well as the design of analytical studies
and the handling of data are every bit as sophisticated as the
preparation of monoclonal antibodies. It is critical that when the
data from immunoassays are to be used for important decisions, that
well designed assays are performed by trained analytical chemists
(3).

When is Immunochemistry Most Applicable?

As indicated above it is very important that as advocates of this
technology, we point out when it is best applied and also when it
should not be applied. This topic has been covered in a variety of
previous reviews (1,4-8), however we have found two figures which
convey several concepts about applicability very well.
        For instance in Figure 1 we represent all of the compounds
for which the analyst may need methods. For some compounds such as
the volatile organics in water, gas chromatography systems offer
great advantages. At the other extreme there are compounds such as
paraquat, the sulfonylureas or benzoylphenylureas which lend
themselves wonderfully to immunoassay development. There is an
important set of compounds which can be analyzed readily by several
different methods. The thiocarbamates or triazines are compounds
handled in this laboratory which fall into such a situation. Here
the decision on the technology would depend upon the resources of
the laboratory in question and the problem at hand. If the
compounds were part of a multianalyte problem or if only a few
samples needed to be assayed, chromatographic systems offer an
advantage. In cases where field assays are needed or where a large
sample load is likely, immunoassay clearly is the method of choice.
        The most common question from biotechnology companies and
from the agricultural chemical industry concerns which compounds

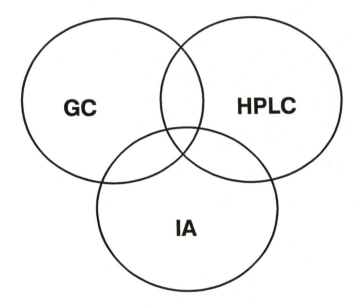

Figure 1. Applicability of immunochemistry to analytical problems. The background indicates all compounds for which analysis is needed while the respective circles indicate the subset of compounds for which gas chromatography (GC), high performance liquid chromatography (HPLC), and immunoassay (IA) are most applicable. For those compounds which can be readily analyzed by a variety of methods, the decision of which assay to use should be made based on the analytical questions to be answered.

are appropriate targets for immunoassay development. Part of this answer of course involves market analysis, which is not an appropriate topic here. It is clear that there will be a reasonable market for immunodiagnostics in the environmental field, but large obvious markets do not now exist.

With the management of the agricultural chemical industry in the past the error has been not to ask what are the proper targets. Rather immunochemical technology is ignored until other analytical methods have failed, the chemistry of the compounds has become cold, and there is tremendous pressure from marketing and registration groups for immediate analytical methods for a very difficult compound. Only then is the task of developing an immunoassay given to a new employee with few resources. This hardly represents the optimum way for a company to develop in-house expertise in immunoassay.

Immunoassays are very versatile, and if one could select but a single method, it could be the method of choice. Fortunately we have a variety of techniques available and a good analyst should know when to apply them. Table I provides some general rules for determining how difficult an immunoassay will be. The terms used are relative and possibly other dimensions to the table could be the laboratory's experience with immunoassay and the problems faced. This table does not indicate that good assays cannot be developed for hard compounds; it just indicates that the expense, skill and time required may be greater for those compounds. For instance we have developed successful immunoassays for some lipophilic, small, unstable, volatile compounds. However, such compounds would be a poor choice to use for one's first venture into immunoassay development.

Table I. Properties of Compounds which Lead to Difficulties in Immunoassay Development

PROPERTIES	
EASY	HARD
HYDROPHILIC	LIPOPHILIC
LARGE	SMALL
STABLE	UNSTABLE
NONVOLATILE	VOLATILE
FOREIGN	NATURAL

Gaining Full Use of Immunochemistry

For over a decade the data have been in the literature to support the contention that for appropriate molecules and problems immunochemical methods are far superior to competing technologies.

Yet, the methods still are not in routine use. Advocates of the technology have entertained themselves by making one immunoassay after another. Although this activity is important, we now face the more difficult challenge of validating these assays, ensuring that they are in the proper hands, and that they are used effectively. This challenge can be broken down into a number of smaller problems which do not differ greatly from problems faced with chromatographic systems. However, some of the problems will still be difficult to address. Fortunately other problems which initially will seem difficult will turn out to be no problem at all.

<u>Should Immunoassays be Qualitative or Quantitative</u>?   This is an excellent example of a nonquestion which sometimes is discussed seriously. As discussed below, the answer is that once one has an antibody and tracer the assay can be put into either a quantitative or qualitative format depending upon the question to be addressed. Qualitative formats will be very important in the environmental arena as fast field tests. However, it is our opinion that at least until the technology is well established that qualitative tests in the environmental field should be based on reagents which have been examined in a quantitative format. Users of qualitative kits which have no quantitative data supporting them could be very embarrassed if they try to over interpret their data.

<u>What Format Should Be Used</u>?   A great strength of immunoassay is that the same reagents can be used in many formats. We have employed Voller's ELISA format (<u>9</u>), but even this format has numerous variations. The format gives adequate sensitivity for most environmental questions, does not require radioactive compounds, can be optimized for speed, cost, sensitivity or other factors, and maybe most important, it has a pleasing and nonintimidating name. In addition we have advocated this format since the understanding that the same antibody can be used in numerous formats is not widespread. We feel that currently it is important to not frighten new users and regulatory agencies with formats for which they have no name recognition.

However, in the long term, ELISA is an ephemeral format. Even when streamlined and automated, it has too many steps. Certainly we should realize that it will be replaced by other systems, the most exciting of which will be biosensors. Also, other formats offer a proprietary edge in the market place which will be very important in the maturation of immunoassay systems in the environmental field. Finally, different formats will lend themselves to different environmental problems. We should continually emphasize that the same reagents can be used in many formats. Possibly in small letters we also should caution that certain antibody characteristics may be more important in one format than another, that some formats are more resistant to matrix effects, and that relative cross reactivities of compounds can change as one changes the subtle principles upon which an immunoassay works. For this reason a clear choice of formats should be made before initiating validation studies.

Clearly, ELISA is the principle format used for introducing immunoassay into the environmental field. We certainly hope that in the near future that all assays will be characterized in this format to avoid confusion.

Should the Development of Antibodies and Antigens be in the Open Literature?   This question certainly is open to debate. On the negative side some companies trying to develop a proprietary niche in the environmental market may feel that they need protection and not divulge their coupling strategies and other techniques in assay development. It is very difficult to obtain patents in the area of hapten chemistry since regardless of how sophisticated we feel our individual work is, the coupling procedures are rather obvious. This is not to say that a great deal of skill and even art are not involved in hapten design and coupling, but that the technologies usually appear obvious in the eyes of patent authorities.

Possibly we suffer from an academic bias, but this laboratory strongly advocates that a general outline of immunoassay development (including the origin and characteristics of the antibody as well as the position and chemistry of hapten coupling to the antigen, and tracer or coating antigen) should be available to the user. The presence of such information in the open literature would not jeopardize a company's art since myopic details are not needed for an analyst to predict the characteristics of the assay. By keeping hapten design secret, companies can confuse many users, but their true competitors usually can discern the general methods used from the characteristics of the antibody. We strongly suggest that the time and public money which must be invested for validation studies only be invested for well characterized assays. To do otherwise would be like developing a chromatographic assay without telling the user what type of detector was being used on the gas chromatograph.

Who Should Develop Immunoassays?   The answer to this question is simple - everyone. The more complex question is once these assays are developed how do we get them in the hands of users? Certainly the agricultural chemical industry should be involved in the development of assays for their products. Even if the assays are never used for registration, the assays will save companies money by being used in-house as research tools. In most companies there is such a backlog of residue samples to run that in-house assays to test formulation, plant distribution, process control and many other problems receive low priority. Immunoassays can have a major impact on these problems.

There also is another answer to the question if agricultural chemical companies should develop assays for their own compounds. That answer is that if they do not, someone else will. If the assay is developed in-house, one has control of the characteristics and distribution of the assay and hopefully the assay development is done correctly. If the assay is not done in-house the company will have no control over quality, sensitivity, or other aspects.

Ultimately it may be cost effective for a company to subcontract assay development to a clinical division in-house or to a third party. However, if this is done before the parent company has the in-house expertise to monitor assay development, this can be a very dangerous and expensive process. Often the expertise on the chemistry of the compound class is not transferred and inferior assays are developed at great expense. If it is necessary to develop early assays outside the company, it is important that the assay development is approached as a collaborative project with a group possessing an established record in the development of assays for environmental samples. Involvement of scientists from the clinical field can be very useful since they have decades of accumulated knowledge on assay formatting and development. However, the involvement of scientists with an appreciation of matrix effects, metabolism, and the regulatory questions posed is critical.

Certainly universities and government agencies should be involved in assay development. If an industrial collaboration can be established, one gains tremendous advantages with regard to chemical libraries and expertise. Universities and government agencies have done an excellent job of pioneering the development of the ELISA technology in the environmental field, but they have two major limitations. The first could be attributed to avarice, administrative incompetence in the institutions, or petty jealousies among the investigators or agencies. There certainly is little easy money to be made from immunoassays for environmental compounds in the near future. Universities and government agencies need to have a policy of providing the assays at no or low cost for research or regulatory use and some fair and systematic method of getting the reagents to third party vendors. As an example, it has taken over a dozen years of pressure from this and other laboratories before the University of California has begun the development of streamlined licensing procedures for immunochemical reagents. Hopefully this problem is being solved in other institutions.

The other problem with university and government laboratories is that they lack expertise in the variety of sophisticated formats which will be very useful in the environmental area and the methods for stabilizing and distributing reagents. Not only should these agencies provide the reagents in a standard format to interested scientists, but by providing them at a reasonable cost to the third party vendors the environmental field will gain their expertise in stabilization, packaging, formatting and marketing the assays. Which company offers the best system then can be determined in the market place.

Clearly biotechnology companies (third party vendors) should be developing kits and in some cases the assays themselves. Hopefully they can get access to the reagents available from government and academic laboratories in addition to the assays developed in-house. As discussed above, we strongly feel that it will be a good policy for these companies to quote the source of antibodies used in their kits or to provide an overview of how the reagents were developed.

Should Monoclonal or Polyclonal Antibodies be Advocated? This subject will be treated elsewhere in this book (10-11) and has been discussed in numerous previous reviews. To most people with experience in the field, this is another nonissue where the answer will be based upon the problem at hand and the resources available. The answer to the question should not be based on the idea that monoclonal antibodies come from high technology and polyclonal from low technology. The sophistication and skill in antibody development can be just as great with either technology. The criteria for approval of a particular assay should be based on rigorous performance specifications of the final product (whether from a commercial or academic source), rather than the design of the test, an approach similar to that used in the manufacture of chromatographic columns. This way of addressing test performance renders most of the questions of antibody selection or standardization moot.

Another misconception is that monoclonals provide an unlimited antibody supply from immortal cells. Hybridoma lines are immortal only so long as they are maintained with constant selection by a skilled technician or frozen in a situation where they can be archived, maintained, and then thawed by a skilled individual. In practice the AOAC sees no difference between the validation of a pool of mono- or polyclonal antibodies used for immunoassay. Also reputable immunochemical companies treat mono- and polyclonal antibodies the same. A sufficient pool of monoclonal or polyclonal antibody is produced and stored such that the company will not have to thaw the hybridoma cells or reimmunize animals in the foreseeable future.

A serious error involves the attempt to use expensive hybridoma screening to overcome poor hapten design and handle recognition. If one is going to the expense of monoclonal production, certainly a similar investment in hapten design to reduce handle recognition is warranted. It is poor economy to use thousands of dollars of hybridoma technology to make up for the lack of a few hundred dollars of hapten design and synthesis.

Both mono- and polyclonal antibodies have a major role and, we will see the role of monoclonal antibodies expanding. For most problems, polyclonal sera will provide adequate sensitivity and specificity faster and at a fraction of the cost of monoclonal antibodies. The idea that any monoclonal antibody will provide greater sensitivity and specificity than a polyclonal is not correct. If one is to invest in monoclonal technology it should be used to develop a large library to the hapten of choice. This library can then be screened to obtain truly superior antibodies for defined applications. For instance one can screen the library for antibodies of high specificity or antibodies which may be class specific. One also could screen the library for antibodies which will give high sensitivity or even in some cases lower sensitivity. Once antibodies are found which give optimum specificity and sensitivity, one can rescreen for antibodies which are resistant to solvents and/or matrix effects. With proper hapten design and a large library, one can screen for the antibodies of the desired characteristics. With the development of biosensors, the availability of defined monoclonal antibodies of varying affinity and avidity will be very important. Unfortunately many hybridoma projects end with scientists finding the antibody they screened for but not the antibody that they wanted.

Although expensive in dollars and time, the investment needed for superior monoclonal antibodies is dropping. This cost may seem high initially, but it is a small investment compared to the major investment needed to characterize and validate an assay. The cost is even small compared to a modern chromatograph and work station. For many compounds this investment will be very cost effective so long as the plan is to obtain a library of superior monoclonal antibodies rather than any monoclonal antibody. Once a monoclonal antibody exists, the cDNAs coding for the respective light and heavy chains can be cloned. These cDNAs then can be engineered to provide very inexpensive antibodies which can be further tailored for applications in immunoaffinity chromatography or biosensor development. Although this added investment seems very high at this time, the technologies involved are advancing rapidly and recombinant antibodies can be anticipated to have a future role in the immunodiagnostic area (12-14). There is even the hope that one may be able to screen for antibodies in bacteria by using recombinant DNA technology (15). Development of these technologies is in the future. However, it is obvious that the field of antibody production is in for some exciting changes. Based on this potential we are placing a major effort in the area of antibody engineering.

How Should Immunoassays for Environmental Samples be Standardized? This question can be broken down into many subtopics relative to good laboratory practice, assay criteria, specifications for immunoassay readers and many more. Obviously the need for standardization will vary depending upon the uses of the assay. Also different regulatory agencies will develop differing criteria. Initially a target could be to use the criteria set forward by the AOAC and discussed in part by Hinton et al. (16) and others (17-20) at this meeting. As discussed below, if some working papers appear on standardization or a committee could be established to provide advice on standardization it would streamline acceptance by not requiring each agency to rediscover the criteria which are useful for acceptance.

In general immunoassay is not hardware intensive. However, the poor reproducibility of binding to some ELISA plates, is a recurring nightmare to analysts. While sources of intraplate variability other than the plates themselves (washing, pipetting error, thermal gradients) may contribute significantly, major differences in variability among plates have been documented (21). One of our studies (22) has identified interwell variability to be by far the largest source of variability. This variability is analogous to chromatographic baseline noise, so it is a critical determinant of assay performance in microplate systems. Also, dimensional standardization is at present a distant dream. In our experience, at least 6 different sets of dimensional specifications are used by the few largest manufacturers of plates. Manufacturers of readers who do not make matching plates must then compromise their specifications to be able to read all of the plates on the market. As with the equipment for chromatographic systems, not all readers are identical in performance (23). It is critical that users look carefully at the specifications of the equipment purchased and have a routine system of rechecking instrument performance. It would be useful to have a committee to make manufacturers of plates and readers fully aware of the unique demands of rigorously quantitative microplate methods. This would hopefully lead to the setting of dimensional and quality standards for plates and readers. These changes could have a dramatic effect on speeding acceptance of the technology and thus expanding the market for enzyme immunoassay (EIA) plates.

What Can Industries and Regulatory Agencies Do to Advance the Technology? A major contribution that these groups can make to the advancement of the technology is to develop the in-house expertise to evaluate the strengths and limitations of the technology. As discussed above a major threat to the technology comes when it is advocated for inappropriate applications.

In the chemical industry the best way to advance the technology is to have an in-house success. This can be accomplished by selecting a chemically reasonable target and planning ahead to obtain adequate chemical support. As mentioned above, it may not be good to select a new product where there will be a great deal of time pressure on the new assay.

Agencies especially can provide a leadership role in several ways. For example the role played by the California Department of Food and Agriculture has been very positive (17), and hopefully other agencies with responsibilities at the national and international levels will take active roles as well. An important contribution is to develop strategic plans for the development of the technology and then attempt to fund work which does not lead to duplication of effort. The private sector will be greatly encouraged if agencies can provide clear procedures for the validation of assays and clear requirements for the data needed.

The most significant role that government could play is in the area of assay standardization; certainly a very active role is possible. A procedure now used by the Food and Drug Administration (FDA) could be immediately implemented by the Environmental Protection Agency (EPA). This would simply involve testing the claims of a manufacturer with regard to the specifications of their particular assay. Any leadership that agencies can provide will benefit the field greatly, and the current effort of the Las Vegas EPA laboratory will have a major impact in this area.

As discussed above getting assays into the hands of users is a major goal. This sometimes is seen as a major hurdle that is different from classical chromatographic methods. The view is that reagents may someday vanish and the assay cannot be performed.

Actually the same situation exists in the chromatographic area. For instance very few laboratories are capable of building their own gas chromatograph or mass spectrophotometer. While there is no guarantee by industry that such equipment always will be available, the market place provides an incentive for third party companies to provide such equipment. The same situation now exists with EIA readers. If regulatory agencies would suggest a system where a pool of antibody and hapten tracer is provided to them for archiving or to a large chemical or biochemical supply company, this fear regarding the availability of reagents might vanish.

A major difference between immunoassay development and the development of a chromatographic assay is that for the former a single moderate investment is needed to develop antibodies and tracers. Adapting the resulting assays to hundreds of laboratories then is relatively cheap. However, with chromatographic assays the developer can assume a heavy investment in equipment in individual user laboratories. Although in some cases the initial cost of assay development may be a little less for a chromatographic system, the total cost to society is dramatically reduced if immunoassays are developed. If government agencies can fund the initial development of a variety of assays or make the development of such assays attractive to third party companies, the rate of acceptance of the technology will increase dramatically.

How Can We Avoid "Turf Wars" in the Immunoassay Field? A major problem that the technology has faced over the last decade has been that there were too few assays. It has been difficult to justify the amount of time needed to learn the technology to analytical laboratories when there are so few applications. We still are in a situation where far too few assays exist. Certainly over the next few years additional groups entering the field will be of great assistance, and we soon will be to the point where enough assays exist for it to be attractive for a residue laboratory to devote a component of its resources to immunochemical analysis.

In spite of the numerous projects in need of scientists working on them, the situation seems to be evolving where several laboratories are working on the same compounds. With some major problems such as the dioxins and dibenzofurans or triazines this clearly is justified. The variety of isomers and metabolites which need to be analyzed as well as the political importance of the class of compounds require the input of several laboratories. In addition a common group of compounds targeted by several laboratories will facilitate comparison of differing technologies. With other projects the resources could be better utilized without duplication, but at this stage of development in the technology, it certainly helps the technology to have procedures repeated independently in several laboratories. This situation clearly is no different from classical methodology where hundreds of "new" analytical methods have been published for DDT, but it is a situation where we need to avoid nonproductive duplication.

Some duplication can be avoided by the agencies that fund the research. Those of us who run soft money laboratories often are in the position of providing the assays for which we receive funds. When different agencies need the same assay, unless there is a major effort at coordination, several laboratories may receive funds to develop assays for the same compound. The chemical industry also should realize that if a regulatory agency does not have access to their assays, that they will have to fund development of a duplicate assay separately. When the chemical industry realizes that it is to their benefit to make assays for their compounds available, then there will be few cases of academic and government laboratories developing assays which already exist in industry. Immunochemical companies who sell qualitative kits or assays where the methods used for assay production are not available to the public should realize again that regulatory agencies or government laboratories may have to develop competing assays.

A burden certainly will fall on academic laboratories for the next few years to ensure that a spirit of collaboration exists among the laboratories in the field. The questions is not who develops the 'first' assay for a compound or the 'best' assay but rather that the field advances and assays get into the hands of users. The technology is complex enough that the field will benefit from different methods even on the same compound. Hopefully with widely used assays such as those for the triazines the laboratories involved will exchange haptens and antibodies and jointly use a library of reagents to generate the multilaboratory data needed for validation.

All laboratories now have the obligation, not only to develop assays, but to get the assays into the hands of users. Rather than racing to develop new assays, possibly we should judge our success based on our ability to transfer the technology successfully to user laboratories. We routinely send reagents to other laboratories in the field. We try to send these reagents with a detailed protocol as well. Not that this is the best way to run the assay, but it represents a method that will give reliable results in the hands of both experts and neophytes.

There is a Need for a Committee to Coordinate the Development of Immunoassays in the Environmental Field. A committee such as the one outlined above now exists in Europe. In the following paragraphs we suggest that such a committee may be of benefit in the United States. However, there is a caution that the committee comes with a variety of problems. For the purpose of this paper the acronym for this Committee for the Evaluation of Immunoassay in Environmental Chemistry will be the palindrome CEIEC.

Possible Roles of CEIEC. The major role of CEIEC would be to act as a cautious advocate for the overall technology rather than a single assay. It also could act as a clearinghouse for information and people dealing with immunoassay and a way for United States researchers to coordinate with scientists internationally. A major goal would be to serve as a forum for discussion of problems relevant to the entire field. Such a committee could encourage investment in the field in general from both the public and private sector. Possibly the most urgent role for such a committee would be to facilitate and coordinate validation efforts for the technology in general as well as for specific assays.

Problems with CEIEC.   In advocating the establishment of such a committee one must consider that it can have both real and perceived problems. CEIEC could be seen as an unfair advocate of one technology or company over another. There is a danger that some regulatory agencies would see it as usurping their roles. This would have to be a very carefully drawn line. The committee could act faster than an agency since its actions would not be legally binding, but the committee's life could be very short if it was seen as a threat to existing agencies. Certainly the committee would have to avoid the charge that it was raising funds or advancing the reputation of one laboratory at the expense of another. Clearly the goal of CEIEC would be to expand rather than to restrict representation which could lead to an unwieldy organization. Its management would present a political tight rope where policies considered good for the field would have to be enforced by mutual acceptance rather than regulation. In this light the committee would be similar to some of the industrial groups trying to avoid pesticide resistance problems. Finally, as a scientific community we must ask if the benefits of such an organization will truly outweigh the added administrative load and even potential dangers should the committee be run in a negative way.

Recent Work in the Immunoassay Area

The 198th ACS meeting certainly is a landmark meeting in the immunoassay field. For the first time at this meeting we have seen reports from a variety of major agricultural chemical companies about the in-house efforts in immunodiagnostics (11,24-29) as well as collaborative validation studies between a biotechnology company and a university (30) and a contract laboratory (31). In addition to the development of polyclonal based systems, there is an increased interest in the development of monoclonal antibodies for environmental chemicals (10,11,32). Deschamps and Hall (32) presented a nice comparison of the relative attributes of mono- vs polyclonal based systems for the herbicide picloram.
    It is reassuring in the Agrochemical Division to see presentation of results on veterinary drugs (27) and environmental compounds which are not pesticides as well as to see the entrance of synthetic chemists into the area (33). As the targets selected for immunoassay development become more difficult, chemical expertise in hapten design will become more critical. This meeting was notable for the first report of the use of computer aided design in analysis of hapten presentation (10). Different animals often have completely different antibody combining sites to the same antigen. Thus, one can anticipate an element of art and luck in hapten design. However, one can stack the odds on the side of a

favorable assay by the application of careful hapten design. A nice example of clever hapten design was presented by Mei and Yin (34) on coupling the carboxylic acid of methoprene. In general in the juvenile hormone field the simplistic approach of attaching the acid directly to a lysine has been used. This work used 4-hydroxybutanoic acid to minimize handle recognition. A similar approach was used with alachlor where a sulfur was used to mimic a chlorine (25). This assay illustrates both a strength and limitation of immunoassay. The hapten design indicates that the assay will detect some major degradation products of alachlor in addition to the parent. Since these workers have described how their assay was made, the characteristic can be used to advantage in exposure studies (29) or the interfering materials easily removed.

The Dupont work (24,28) in addition to several other studies provides an excellent correlation between immunochemical and classical methods. The work also provides a useful caution that with such sensitive assays extreme care is needed in sample handling. Excellent correlations also were obtained between classical and immunochemical methods with clomazone (26). A caution common to both assays is that neither correlates perfectly with bioassay. This is a reminder that immunoassays are physical assays with no magical biological properties. An interesting validation study using triazine antibodies indicated that high pressure liquid chromatography (HPLC) detected atrazine in a sample while immunoassay did not. When the sample was further analyzed by gas chromatography-mass spectrometry, the ELISA results were confirmed (11). This certainly does not indicate that ELISA is superior to HPLC, but that the different methods complement each other and can be used to cross check each other.

As with our work with a variety of compounds, the work from Dupont on triazines (28) illustrates that the same antibody can give assays of vastly different specificity and sensitivity if it is used with a different coating antigen. This again illustrates the importance of a laboratory developing a library of antibodies and antigens for a whole class of compounds and the respective metabolites, rather than a piece meal approach to assay development.

Several of the papers presented demonstrate that the same antibody can be used in a variety of different formats (31-32). This characteristic will become increasingly important. Certainly the same assay can be used both for analysis of environmental samples and in the analysis of human body fluids as a biomarker approach (29,35); in the latter application immunoassay offers numerous advantages.

Many of the problems now faced by immunoassay mentioned above and in other articles (3,4,36) clearly are being addressed by scientists in academic, governmental and industrial laboratories. With the level of expertise that is now evident in the field, one can have confidence that the problems will be solved and that immunochemical assays will assume their rightful role as one of the tools of the modern environmental chemist.

Representative Work from This Laboratory

Assay development can be described in a timeline such as shown in Figure 2. We will illustrate points on this timeline with examples from our laboratory.

An example of an assay developed in our laboratory which is well along the timeline is the one for the herbicide molinate. This compound is relatively volatile and has a relatively hydrolytically unstable thiocarbamate bond. Four haptens were synthesized by a thiol replacement reaction with thiocarbamate sulfones which left the hexahydroazepine ring unmodified. Two of the haptens had alkyl chain spacers terminating in a carboxylic acid. The other two had p-aminophenyl spacers. Antibodies against an alkyl chain derivative conjugated to keyhole limpet hemocyanin (KLH) were used in an indirect competitive ELISA format with a p-aminophenyl hapten conjugated to conalbumin as the coating antigen. This assay had a limit of detectability of about 3 ppb and an I_{50} of approximately 100 ppb. A laboratory dissipation study was conducted and samples analyzed by liquid scintillation counting and ELISA. Samples were either added directly to liquid scintillation cocktail and counted or diluted in buffer and mixed with antibody for the ELISA determination. This pilot study confirmed that the ELISA could quantitatively measure molinate in samples, with the advantage of not needing extraction prior to analysis. Details of the hapten synthesis, assay development and optimization were reported by Gee et al. (37).

To further validate the assay for use with environmental samples, water samples spiked with molinate were extracted and analyzed by ELISA and GC. Recovery comparisons were made between ELISA and GC for both liquid-liquid and solid phase extraction methods. Recoveries were greater than 90% for levels as low as 1ppb for all analysis and extraction method comparisons (38). This study also described the utility and compatibility between solid phase extraction and ELISA for measuring low concentrations of molinate. As much as 10% acetonitrile/propylene glycol (1:1) or 5% methanol had no effect on the molinate assay. Details of this study were reported by Li et al. (38).

Subsequently we have completed an extensive validation study using field samples which contained high concentrations of molinate following an aerial application. These samples were analyzed by ELISA directly after buffering and confirmed by GC analysis of split samples. One of the most valuable lessons from this validation study was the importance of the various quality control considerations (22). From sigmoidal standard curves, 20-60 percent of the control absorbance was determined experimentally to be the region of greatest precision. Thus sample concentrations arising from data generated outside this area would be less reliable. Control charts were constructed for both positive and negative control samples as a means of evaluating assay performance over the study period. Such charts can be useful indicators of changes in the assay that may affect reported results and are a commonly used tool in clinical chemistry. These control data were also run through a nested analysis of variance. The largest relative error contributions arose from well to well (or well replicate) variability. Other details of this study such as data handling and other sources of procedural error can be found in Harrison et al. (22).

We view this series of studies as an essential prototype for the entire development and validation process. For example, we have also completed a similar study for molinate using an improved format to analyze low concentration samples obtained from the Sacramento River and associated drainage canals. We are also using this prototype in the development and validation of our assays for triazines (39). Our experience in the validation of the molinate assay, especially our understanding of the quality control problems, has been crucial to our successful transfer of the triazine assays to the California Department of Food and Agriculture (CDFA) and other laboratories for routine application to environmental samples.

The same assay which was transferred to CDFA has been used by our laboratory to demonstrate the usefulness of immunoassay for screening water samples. In this study, 75 well water samples were analyzed by GC and immunoassay for triazines. The background level of the immunoassay was 15 ppt, determined by repeated solid phase extraction and analysis of reagent water blanks; the highest level detected was approximately 0.3 ppb. The coefficient of variation for a single sample run 6 times was 10%. The variability of the two methods was comparable based on analysis of 18 paired samples; the mean coefficient of variation was 11% for the ELISA and 13% for the GC. The most valuable attribute of this application is the low false negative rate. None of the 40 samples having the lowest concentrations by ELISA were positive by GC. This assay is now being used in a large scale field test by CDFA as well as in a worker exposure study.

An important extension of our large validation studies involves the use of data bases from field studies in the development of improved statistical methods for a variety of problems in quantitative applications of immunoassays. These problems include the preparation and analysis of calibration curves, treatment of "outliers" and values below detection limits, and the optimization of resource allocation in the analytical procedure. This last area is a difficult one because of the multiple level nested designs frequently used in large studies such as ours (22). We have developed collaborations with David Rocke and Davis Bunch (statisticians and numerical analysts at Davis) in order to address these problems within the context of working assays. Hopefully we also can address the mathematical basis of using multiple immunoassays as biochemical "tasters" to approach multianalyte situations.

As mentioned above and in various reviews (1,4,6-8), hapten synthesis is the first and probably one of the most important steps in assay development. The most general "rules" in hapten design are to locate the spacer attachment distal to important haptenic determinants to maximize their exposure for antibody binding; spacers containing strong determinant groups should be avoided to minimize the production of spacer specific antibodies; functional groups used for coupling must be compatible with target molecule functional groups to avoid cross-linking or modifying the target during conjugation; consideration of hapten and target stability under conjugation, immunization and assay conditions; selection of hapten to improve solubility or at least avoid solubility problems and to minimize the number of synthetic steps by using commerically available materials or, in some cases, direct coupling to the target. We have described and examined these basic criteria for hapten synthesis in some detail using examples from our laboratory and the literature (40).

The significance of these criteria is demonstrated routinely in our laboratory. In some cases, however, where the development of an immunoassay may be difficult, the choice of the hapten to be synthesized may depend greatly on the ultimate use of the assay with samples. For example, bentazon, a rice herbicide, is a small molecular weight (MW 240) molecule with an unique acidic secondary sulfonamide (pKa 3.4). An N-derivatized bentazon compound was synthesized. This changed the molecule from an acidic secondary sulfonamide to a neutral tertiary sulfonamide. Antisera raised against this N-derivative conjugated to KLH showed 2-orders of magnitude greater binding to N-ethyl bentazon than to bentazon (Figure 3). An assay such as this could be useful in assessing bentazon concentrations after the sample has been ethylated. Sample derivatization prior to GC analysis is a commonly used technique. With compounds for which immunoassay development is difficult, due to the presence of multiple reactive groups, antibodies against derivatized compounds is an alternative.

Not all of our assay development work is successful and we have found it instructive to analyze our negative results in some detail. This is especially important for failures of hapten design, where no useable antibodies to the target compound are obtained. There are a number of strategies for attaching the hapten to a carrier molecule. One is to attach the spacer arm to the protein, then attach the hapten to the free functional group of the spacer arm (41). We have found that this conjugation strategy failed to produce high affinity antibodies for both amitrole (Figure 4) and bentazon, yielding instead antibodies which primarily recognize the spacer. Similar data have been obtained by others (42-43). These examples emphasize the value of the approach to antibody screening described by Harrison et al. (40) in understanding negative data.

In optimizing an assay during development the nature of the interaction of the analyte with the antibody is particularly important. Assays usually are carried out under physiological conditions and frequently no effort is made to optimize for pH, ionic strength, or other factors. These factors can directly affect the assay by modifying the presentation of the soluble analyte to the antibody or changing the interaction of the antibody and the conjugated hapten used in the assay. For example, assays for some compounds show a distinct pH dependence. In an indirect competitive

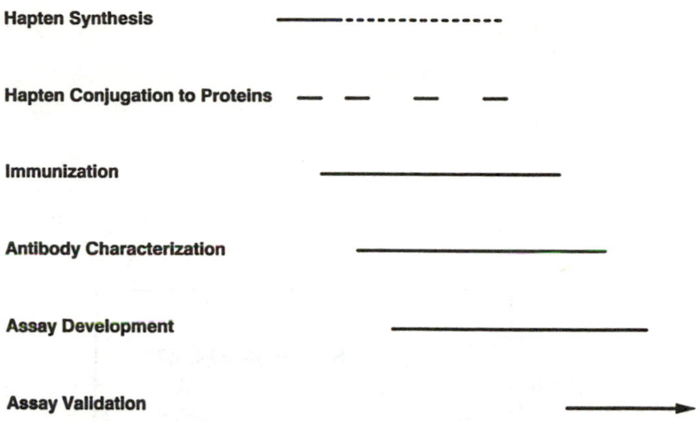

Hapten Synthesis ———————-------------

Hapten Conjugation to Proteins — — — —

Immunization ————————————

Antibody Characterization ————————————

Assay Development ————————————

Assay Validation ——————————————➤

Figure 2. Timeline illustrating the relationship among the various assay development and implementation steps. It is critical that hapten preparation occur first. However existing assays can be improved by rational improvements in reagents or format. Once a validation study is undertaken, it is important to use a constant format and reagent set.

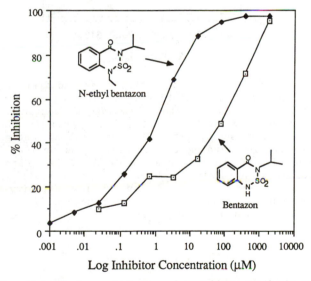

Figure 3 . Relative sensitivity of a rabbit antibody to bentazon and methylated bentazon. A rabbit antisera against an N-derivatized bentazon had better recognition of methylated bentazon than bentazon. Coating antigens were Bz(6)-O-MPAA-BSA and Bz-succ-BSA for methyl bentazon and bentazon respectively. These curves indicate that one may find a much more sensitive assay for a derivative than for the parent compound. As in chromatographic analysis, it may be advantageous to run immunoassays on derivatives and an immunoassay such as this could easily be used to quantitate derivatized bentazon samples.

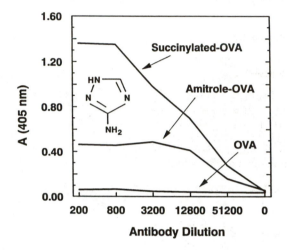

Figure 4. Rabbit antibody specificity for different coating antigens after the second bleeding. Binding of the anti-aminotriazole antibody to the homologous coating antigen, amitrole-succinylated ovalbumin (OVA), compared to the native protein, OVA, and succinylated OVA as coating antigens. The antibodies show low binding to the homologous antigen but high recognition of the succinylated protein. These data show a common problem when raising antibodies to very small molecules. The antibodies from this bleed have a low affinity for the aminotriazole hapten while the hemisuccinate used as a spacer on the succinylated protein is antigenic.

ELISA, the sensitivity of the assay for the herbicide glyphosate was improved by at least one order of magnitude when the assay was conducted at pH 5.8, rather than 7.3 (Figure 5). Glyphosate (phosphonomethylglycine) has several Zwitterionic forms, so it is not surprising that careful optimization of pH led to a dramatic assay improvment as did a shift away from phosphate buffers. This work by Dr. Jung is a dramatic demonstration of how a series of optimizations can improve the sensitivity of assays several orders of magnatude. Along the same line, Sharp et al. (24) have reported that 0.01M CaCl$_2$ greatly improves the sensitivity of some, but not all assays for chlorsulfuron. Although optimization of sensitivity is important, it is also necessary to recognize that there may be a tradeoff between the increased sensitivity and assay ruggedness.

Figure 6 shows the class recognition of one of the triazine antibodies produced in our laboratory. Immunoassays for the triazines will be very interesting due to the existance of numerous structural analogs in this important class of herbicides. Although most degradation products lack herbicidal activity, they can be important analytical targets as indicators of human or environmental exposure. The antibody in Figure 6 recognizes triazines having a -Cl or -SCH$_3$ in the 2 position of the ring, such as atrazine (I$_{50}$ = 6.5 ppb), simazine (IC$_{50}$ = 54 ppb) and ametryne (I$_{50}$ = 130 ppb), regardless of minor changes in the N-alkyl substitution pattern. The monodealkylated or 2-hydroxy metabolites are also recognized, though to a lesser degree (I$_{50}$ > 3500 ppb). We have obtained similar results for several other rabbit antibodies and five mouse monoclonal antibodies. Such antibodies can be used for direct analysis of triazines by ELISA, separation of related triazine species by immunoaffinity chromatography, or removal of triazines from contaminated samples. The relative recognition of the various triazines and their metabolites depends on the hapten used to produce the antibody. Variables we have explored thus far in our work on the triazines include position of conjugation, spacer length, and alkyl group substitution pattern (39-40). Use of a library of antibodies and coating antigens can result in either class or compound specific assays. A series of related assays can be used to screen samples for certain substitutions, aiding identification of the immunoreacting compounds.

We have also applied ELISA to several biological pesticides including the endotoxin of Bacillus thuringiensis kurstaki (Btk). In this application to a macromolecular analyte, we have used a double antibody sandwich ELISA for Btk to measure the amount of ELISA reactive material in formulations of the pesticide. Figure 7 shows the use of an ELISA standard curve of gel purified Btk endotoxin to measure the immunoreactive material in dilutions of two Btk formulations. It has been demonstrated that ELISA can serve as a quick quality control check for formulations of Bacillus thuringiensis israelensis (44). Such examples indicate that immunoassays will be increasingly important as biologicals and products of recombinant DNA research impact our field (44).

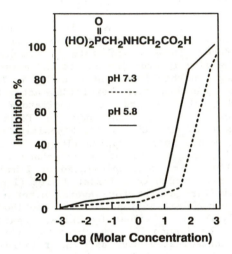

Figure 5. Inhibition of anti-glyphosate antibody by glyphosate. Inhibition curves for polyclonal anti-glyphosate antibodies raised in rabbits were conducted in 50 mM TRIS buffer at pH 5.8 and pH 7.3. The curves show an increased affinity between antibody and glyphosate at the lower pH buffer, illustrating that for some compounds, optimization for pH is critical. Careful optimization of assay conditions is especially important as the molecule becomes smaller, for zwitterionic materials, and for analytes where hydrogen bonding may play a major role in antibody binding.

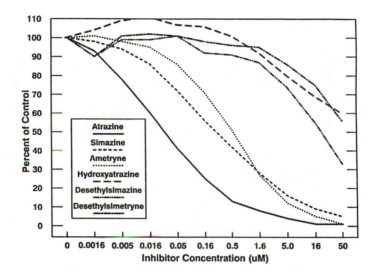

Figure 6. Competitive inhibition ELISA results. Recognition of atrazine and several related compounds, including metabolites, by a rabbit antiserum raised against a conjugate of an atrazine hapten and a carrier protein. This antibody recognizes triazines having either a -Cl or $-SCH_3$ in the 2 position of the ring, such as atrazine and ametryne, regardless of minor changes in the N-alkyl substitution pattern. By careful design of the immunizing and coating antigen, one can vary the properties of the resulting assay to detect a single compound or a predictable set of related compounds.

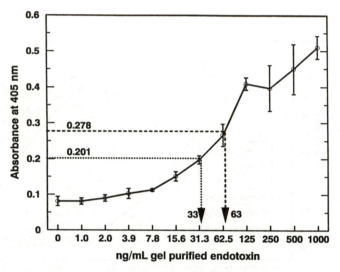

Figure 7. Standard curve of gel purified 60 kd protein endotoxin of Btk was generated using a double antibody sandwich ELISA. The arrows indicate dilutions of two Btk formulations; absorbance values were used to determine the endotoxin concentrations of the formulations, based on the standard curve. The formulation dilutions gave curves that were virtually superimposable on the standard curve. Such similarity in shape and slope indicate that the antibody is likely binding to a specific determinant common to the purified Btk and the Btk in the formulation. In general immunoassays for biopolymers are much easier to develop than assays for small molecules. However, only recently has an interest in trace analysis of such materials begun to develop in the environmental field. Thus, sample cleanup and handling is not as sophisticated as with small molecules.

Summary

Immunoassays are now being seen as useful supplements to classical chromatographic analytical systems. In the near future we will see also an integration between immunochemical and classical procedures as immunochemical methods are used to prioritize or clean up samples before chromatography or as a post-column detection system. If "Green" initiatives in several countries pass we will see two striking trends. The first will be a shift of agricultural production to other areas and the second will be an acceleration in the development of biological methods of pest control. Immunochemistry offers tremendous advantages for inspection of the large increase in imported food and may be the only viable analytical method for many biological pesticides.

Acknowledgments

This work was supported in part by NIEHS Superfund PHS ES04699, EPA Cooperative Agreement No. CR-814709-01-0, and a grant from the California Department of Food and Agriculture. B.D.H. is a Burroughs Wellcome Scholar in Toxicology. K.M.S.S. was on leave from Forestry Canada, Forest Pest Management Institute, Sault Ste. Marie, Ontario Canada. Notice: Although the research described in this article has been supported by the United States Environmental Protection Agency (through assistance agreement CR-814709), it has not been subjected to Agency review and therefore does not necessarily reflect the views of the Agency and no official endorsement should be inferred.

Literature Cited

1. Hammock, B.D.; Mumma, R.O. In *Pesticide Analytical Methodology*; Zweig, G., Ed.; ACS Symposium Series No. 136; American Chemical Society: Washington, DC, 1980; pp 321-352.
2. Langone, J.J.; Van Vunakis, H. *Res. Commun. Chem. Pathol. Pharmacol.* 1975, *10*, 163-171.
3. Hammock, B.D. In *Biotechnology for Crop Protection*; Hedin, P.A.; Menn J.J.; and Hollingworth, R.M., Eds.; ACS Symposium Series 379: American Chemical Society: Washington, DC, 1988; p 298.
4. Van Emon, J.M.; Seiber, J.N.; Hammock, B.D. In *Bioregulators for Pest Control*; Hedin, P., Ed.; ACS Symposium Series No. 276; American Chemical Society: Washington, DC, 1985; pp 307-316.
5. Hammock, B.D.; Gee, S.J.; Cheung, P.Y.K.; Miyamoto, T.; Goodrow, M.H.; Van Emon, J.; Seiber, J.N. In *Pesticide Science and Biotechnology*; Greenhalgh, R. and Roberts, T.R., Eds.; Blackwell: Oxford, 1986; p 309.
6. Harrison, R. O.; Gee, S.J.; Hammock, B.D. In *Biotechnology for Crop Protection*; Hedin, P.A.; Menn J.J.; and Hollingworth,

R.M., Eds.; ACS Symposium Series 379: American Chemical
Society: Washington, DC, 1988; p316.

7. Jung, F.; Gee, S.J.; Harrison, R.O.; Goodrow, M.H.; Karu, A.E.
    Braun, A.L.; Li, Q.X.; and Hammock, B.D. Pestic. Sci. 1989,
    26, 303-317.

8. Van Emon, J.M.; Seiber, J.N.; Hammock, B.D. In Analytical
    Methods for Pesticides and Plant Growth Regulators; Sherma,
    J., ED.; Vol. XVII; Academic Press: New York, 1989; pp 217-
    263.

9. Voller, A.; Bartlett, A.; Bidwell, D.E. J. Clin. Pathol. 1978,
    31, 507-520.

10. Karu, A.E.; Schmidt, D.J. Clarkson, C.E.; Carlson, R.E.; and
    Van Emon, J.M. 198th ACS National Meeting, Division of
    Agrochemicals, Miami Beach, Florida, 1989, paper 9.

11. Schlaeppi, I.-M.; Fory, W.; Ramsteiner, K. 198th ACS National
    Meeting, Division of Agrochemicals, Miami Beach, Florida,
    1989, paper 25.

12. Bird, R. E.; Hardman, K. D.; Jacobson, J. W.; Johnson, S;
    Kaufman, B. M.; Lee, S.-M.; Lee, T.; Pope, S. H.; Riordan, G.
    S.; Whitlow, M. Science 1988, 242, 423-6.

13. Rodwel, J. D. Nature 1989, 342, 99-100.

14. Hiatt, A.; Cafferkey R.; Bowdish, K. Nature 1989, 342, 76-78.

15. Huse, W.D.; Sastry, L.; Iverson, S.A.; Kang, A.S.; Alting-Mees,
    M.; Burton, D.R.; Benkovic, S.J.; Lerner, R.A. Science 1989,
    246, 1275-1281.

16. Hinton, D. M.; Albert, R. H.; Horwitz, W. 198th ACS National
    Meeting, Division of Agrochemicals, Miami Beach, Florida,
    1989, paper 56.

17. Stoddard, P. J. 198th ACS National Meeting, Division of
    Agrochemicals, Miami Beach, Florida, 1989, paper 20.

18. Berkowitz, D. B. 198th ACS National Meeting, Division of
    Agrochemicals, Miami Beach, Florida, 1989, paper 19.

19. Van Emon, J. M. 198th ACS National Meeting, Division of
    Agrochemicals, Miami Beach, Florida, 1989, paper 18.

20. Pohland, A. E. 198th ACS National Meeting, Division of
    Agrochemicals, Miami Beach, Florida, 1989, paper 17.

21. Shekarchi, I.C.; Sever, J.L.; Lee, Y.J.; Castellano, G.;
    Madden, D.L. J. Clin. Microbiol., 1984, 19, 89-96.

22. Harrison, R.O; Braun, A.L.; Gee, S.J.; O'Brien, D.J.; Hammock,
    B.D. Food Agric. Immunol. 1989, 1, 37-51.

23. Harrison, R. O.; Hammock, B. D. J. Assoc. Off. Anal Chem.,
    1988, 71, 981-7.

24. Sharp, J.; Lueng, F.S.; O'Brien, D.P.; Carski, T.H. 198th ACS
    National Meeting, Division of Agrochemicals, Miami Beach,
    Florida, 1989, paper 7.

25. Feng, P.C.C.; Wratten, S.J.; Sharp, C.R.; Horton, S.R.; and
    Logusch, E.W. 198th ACS National Meeting, Division of
    Agrochemicals, Miami Beach, Florida, 1989, paper 8.

26. Dargar, R.V.; Tymonko, J.M. Van Der Werf, P. 198th ACS National
    Meeting, Division of Agrochemicals, Miami Beach, Florida,
    1989, paper 10.

27. Wong, R.B. 198th ACS National Meeting, Division of Agrochemicals, Miami Beach, Florida, 1989, paper 26.
28. Sharp, J.K.; Marxmiller, R.L.; Robotti, K.; Ehrmann, P.; Sereno, R. 198th ACS National Meeting, Division of Agrochemicals, Miami Beach, Florida, 1989, paper 29.
29. Sharp, C.R.; Feng, P.C.C.; Horton, S.R.; Logusch, E.W. 198th ACS National Meeting, Division of Agrochemicals, Miami Beach, Florida, 1989, paper 30.
30. Bushway, R.J.; Ferguson, B.S. 198th ACS National Meeting, Division of Agrochemicals, Miami Beach, Florida, 1989, paper 4.
31. Reck, B.; Frevert, J.; Knoell, H.E. 198th ACS National Meeting, Division of Agrochemicals, Miami Beach, Florida, 1989, paper 31.
32. Deschamps, R.J.A.; Hall, J.C. 198th ACS National Meeting, Division of Agrochemicals, Miami Beach, Florida, 1989, paper 27.
33. Eck, D.L.; Kurth, M.J.; Macmillan, C.B. 198th ACS National Meeting, Division of Agrochemicals, Miami Beach, Florida, 1989, paper 11.
34. Mei, J.V.; Yin, C.-M. 198th ACS National Meeting, Division of Agrochemicals, Miami Beach, Florida, 1989, paper 28.
35. Lindley, K.; White, R.J. Van Emon, J. 198th ACS National Meeting, Division of Agrochemicals, Miami Beach, Florida, 1989, paper 32.
36. Seiber, J.N.; Li, Q.X.; Van Emon J. 198th ACS National Meeting, Division of Agrochemicals, Miami Beach, Florida, 1989, paper 16.
37. Gee, S.J.; Miyamoto, T.; Goodrow, M.H.; Buster, D.; Hammock, B.D. J. Agric. Food Chem. 1988, 36, 863-870.
38. Li, Q.X.; Gee, S.J., McChesney, M.M.; Hammock, B.D.; Seiber, J.N. Anal. Chem. 1989, 61, 819-823.
39. Goodrow, M.H.; Harrison, R.O.; Hammock, B.D. 1990, J. Agri. Food Chem. in press.
40. Harrison, R.O.; Goodrow, M.H.; Gee, S.J.; Hammock, B.D. To be published in an ACS Symposium Series volume based on the symposium Immunoassays for Monitoring Human Exposure to Toxic Chemicals in Food and the Environment; Vanderlaan, M., Ed.; American Chemical Society: Washington, DC.
41. Hung, D.T.; Benner, S.A.; Williams, C.M. J. Biol. Chem. 1980, 255, 6047-6048.
42. Harrison, R.O. Ph.D. Thesis, University of Maryland, College Park, 1987.
43. Jung, F.; Szekacs, A.; Hammock, B.D. To be published in an ACS Symposium Series volume based on the symposium Immunoassays for Monitoring Human Exposure to Toxic Chemicals in Food and the Environment; Vanderlaan, M., Ed.; American Chemical Society: Washington, DC.
44. Cheung, P.Y.K.; Hammock, B.D. In Biotechnology for Crop Protection; Hedin, P.A.; Menn J.J.; and Hollingworth, R.M., Eds.; ACS Symposium Series 379: American Chemical Society: Washington, DC, 1988; 359-372.

RECEIVED July 31, 1990

Chapter 12

An Enzyme-Linked Immunosorbent Assay for Residue Detection of Methoprene

J. V. Mei[1], C.-M. Yin[1], and L. A. Carpino[2]

[1]Department of Entomology and [2]Department of Chemistry, University of Massachusetts, Amherst, MA 01003

An enzyme-linked immunosorbent assay has been developed for the detection of low levels of methoprene from tobacco samples. The generation of anti-methoprene antibodies needed for such an assay relied on the preparation of a methoprene-carrier immunogen. Methoprene acid was covalently bound via an ester to a spacer group, which in turn was bound to a protein carrier. Two activated ester methods were used to prepare the immunogen, one of which forms a water soluble, activated ester of methoprene. The resulting polyclonal antibodies raised against the methoprene immunogen were highly specific for methoprene, and did not cross react with closely related esters such as the insect juvenile hormones. The range of the methoprene ELISA was from 5 to 300 ng/mL, with an I_{50} of 50 ng/mL.

Through the study of the role of hormones in the regulation of insect life processes, insect hormone analogs have been developed and used in insect control programs (1-2). Of special interest to entomologists are those hormones and their analogs that can regulate growth, metamorphosis and reproduction (3). Several types of insect growth regulators (IGRs) have been developed to date which are based on the structure of insect hormones. One of the most thoroughly studied and widely used IGRs is methoprene (4), which mimics the action of juvenile hormone (JH) and disrupts insect metamorphosis. Methoprene (isopropyl 11-methoxy (E,E)-3,7,11-trimethyl-2,4-dodecadienoate) (Figure 1, Structure 1) is used in many commercial formulations for the control of a variety of pests including mosquitoes, flies, ants, fleas, aphids and stored-product pests. As the regulation of pesticides becomes

0097–6156/90/0442–0140$06.00/0

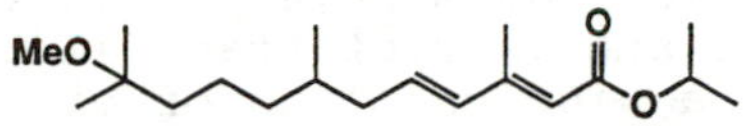

Methoprene

1

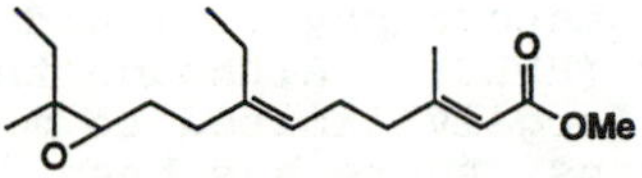

Juvenile Hormone I (JH I)

2

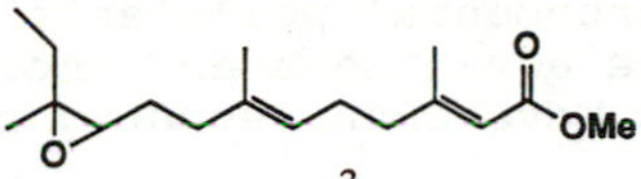

JH II

3

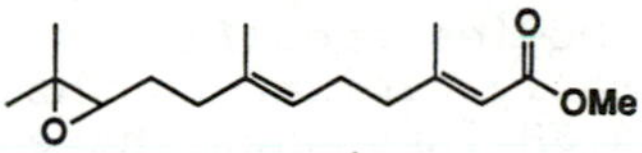

JH III

4

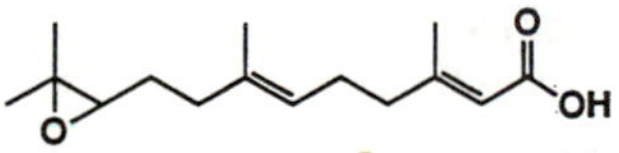

JH III acid

5

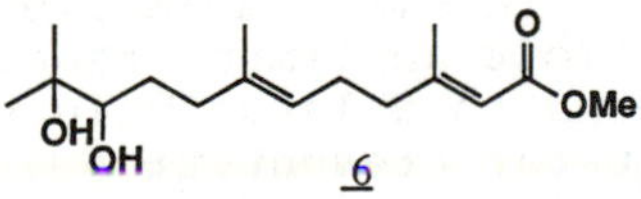

JH III diol

6

Figure 1. The structure of methoprene, the juvenile hormones and JH derivatives.

more restrictive, and with the increased use of methoprene predicted for the future (Edman, J.D.; Clark, J.M. <u>Mosquito Control Board, Massachusetts Department of Food and Agriculture</u>, under review), a need has arisen for the detection of methoprene residues by an easy, economical, sensitive and reliable assay.

Immunochemical techniques have proven to be powerful tools in the analysis of trace amounts of organic compounds in a wide spectrum of biological and environmental matrices. They have been viewed as a tool to supplement the conventional analytical techniques of gas chromatography (GC) and high pressure liquid chromatography (HPLC). Although the use of immunoassay has been largely limited to medical and biochemical applications, there has been increased interest in the use of immunochemical techniques for the detection of environmental pollutants and pesticide residues. Some of the advantages and shortcomings of immunochemical assays have been recently outlined (<u>5-6</u>).

One of the key steps in the development of any immunochemical assay for a small organic compound, is the production of antibodies specific to that compound. Methoprene is structurally similar to the JHs (Figure 1, Structures 2-4) and JH immunogens have been prepared by several groups of researchers. These studies provided the basis for the design and synthesis of a methoprene immunogen (<u>7-9</u>).

Design of the Methoprene Immunogen

Both methoprene and JH are small molecules and by themselves will not elicit an immune response. They must be covalently bound to a large molecule, such as a protein, which facilitates presentation of the small molecule to a mammal's immune system. Such small molecules, when bound to carriers for immunological purposes, are known as haptens.

Two methods of forming JH immunogens were developed. The first method was based on the hydrolysis of the methyl ester of JH (Figure 1, Structure 4) to the free acid (<u>7-8</u>) (Figure 1, Structure 5). The acid was directly conjugated to the protein, human serum albumin, using the N-hydroxysuccinimide active ester. The resulting immunogen lacked the methyl ester function of the native hormone. Antisera raised against this JH acid immunogen recognized both JH and JH acid.

A second method involved the hydrolysis of the JH epoxide (Figure 1, Structure 4) to the corresponding diol (<u>9</u>) (Figure 1, Structure 6). The diol was conjugated to human serum albumin via a succinyl group, which resulted in loss of the epoxide feature of the native hormone. The antiserum raised against this immunogen recognized only the JH diol derivative.

From the above results, it was hypothesized that if the methoxy and ester functions of methoprene are preserved in the immunogen, it may induce the production of antibodies with higher specificity for methoprene. In this work, an immunogen for methoprene (Figure 4, Structure 15) was developed which maintained its native structure by incorporating a spacer group between the methoprene molecule and the protein carrier. A four-carbon spacer group was linked to methoprene as an ester, which in turn, was linked to the protein as an amide. Thus the methoxy function remained untouched and the ester function remained intact. Additionally, work using liposomes to carry haptens into a mammalian system has shown that a spacer group greatly increases the immunogenicity of the hapten (10-11).

The parent spacer, 4-hydroxybutanoic acid, spontaneously undergoes loss of water under acidic or basic conditions to give 4-butyrolactone. To prevent cyclization when the spacer is coupled to methoprene, both the hydroxyl and carboxyl functions had to be protected and selectively deprotected. The methods used are summarized in Figure 2. The protection/deprotection chemistry described was originally developed for use in the field of peptide synthesis.

Immunogen Synthesis

Incorporation of the Spacer Group. 4-Benzyloxybutanoic acid (Figure 3, Structure 7) was prepared (12), using a benzylether group as protection for the hydroxyl function. The protected acid was coupled to 2-trimethylsilylethanol (Figure 3, Structure 8) using dicyclohexylcarbodiimide (DCC) (13), to give compound 9 (Figure 3), in which the trimethylsilylethyl (TMSE) group protects the carboxylic acid function. The benzyl group was next removed using catalytic hydrogenation (15) to provide the protected alcohol (Figure 3, Structure 10).

The TMSE-protected alcohol (Figure 3, Structure 10) was then coupled to methoprene acid (Figure 3, Structure 11) using DCC. Methoprene acid was obtained by base-catalysed hydrolysis of the isopropyl ester (Figure 1, Structure 6). The resulting protected methoprene derivative (Figure 3, Structure 12) was deblocked via tetraethylammonium fluoride (13) to give the desired methoprene-spacer acid (Figure 3, Structure 13). Acid 13 (Figure 3) and all intermediates were

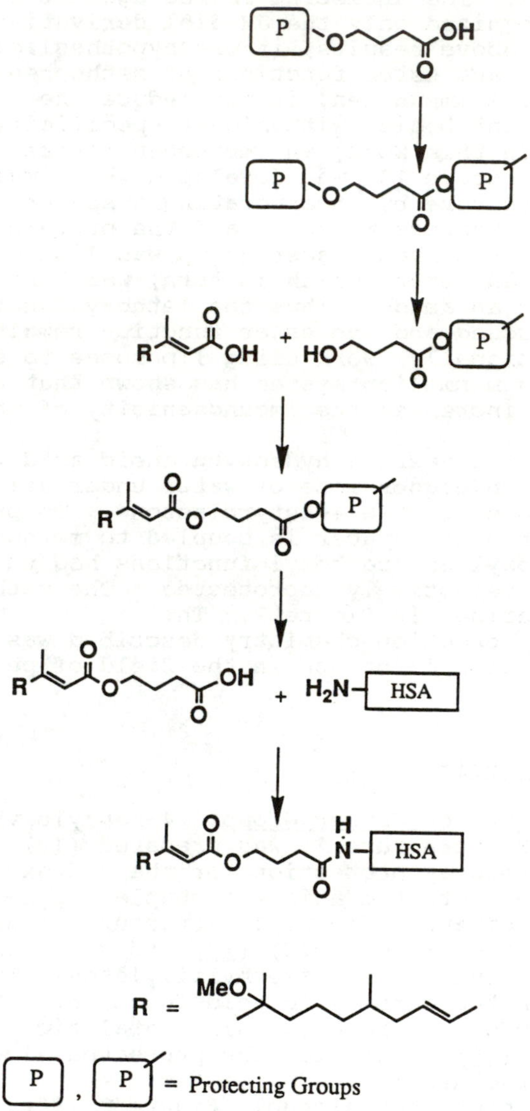

Figure 2. Diagramatic scheme for the protection/deprotection steps required to place a spacer group between a hapten and a carrier.

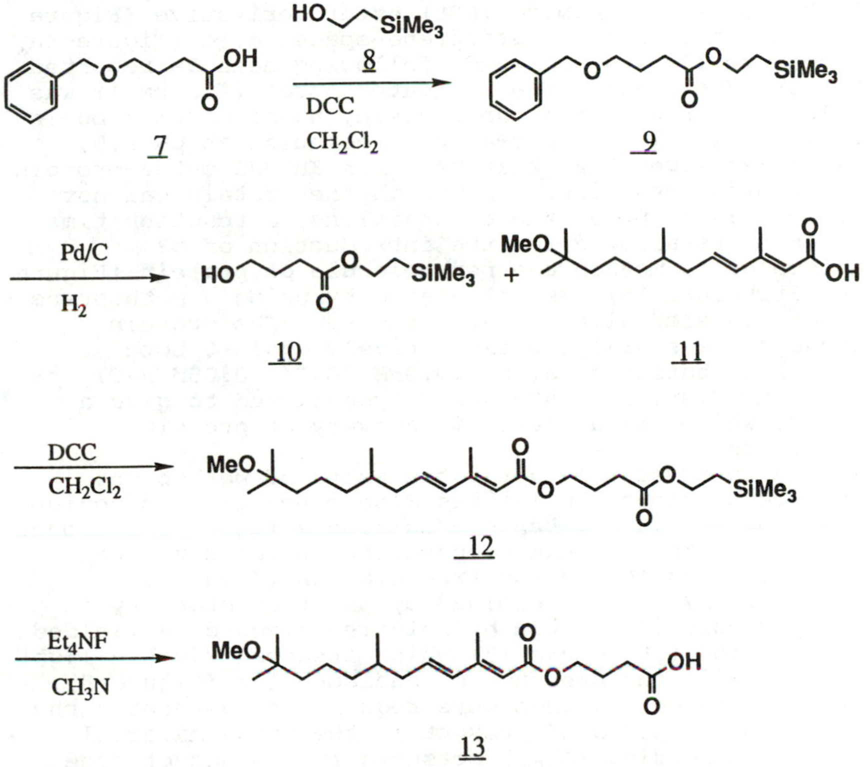

Figure 3. Synthesis of the protected four-carbon spacer group (Structure 9), followed by coupling to methoprene acid (Structure 11) which yielded Structure 12. Deprotection of compound 12 gave methoprene-spacer acid (Structure 13).

fully characterized using NMR and IR spectroscopy and
elemental analysis (15).

Synthesis of Activated Esters. Haptens are frequently
conjugated to carriers by means of activated esters.
An N-hydroxysuccinimide (NHS) ester derivative (Figure
4a, Structure 14) of methoprene-spacer acid (Figure 4a,
Structure 13) was prepared, following methods reported
for the JHs (7-8). The activated ester (9.2 μmol) was
allowed to react with the protein, human serum albumin
(0.17 μmol), in organic-aqueous solution at pH 8.5,
which was lower than that used for JH-NHS ester-protein
conjugation reactions. Although the protein was not
fully soluble under these conditions, a reaction time
of 48 hours allowed for the introduction of 53
molecules of methoprene per molecule of protein (Figure
4a, Structure 15), as calculated by using a methoprene
tracer labeled with ^{14}C at carbon-5. The protein
conjugate was dialysed exhaustively against both
phosphate buffered saline (0.05M PO_4^{-2}, 0.05M NaCl, pH
7.4) and distilled water, and lyophilyzed to give a
fluffy white solid (90-95 % recovery of protein
conjugates).
    A water soluble activated ester of methoprene
(Figure 5, Structure 16) was also prepared from sodium
1-hydroxy-2-nitro-4-benzene sulfonate (16). The amount
of compund 16 in crude preparations which contained
both compound 16 and the free dianion (Figure 5,
Structure 17) was determined by spectrophotometry in
aqueous solution. Upon hydrolysis compound 16 yielded
the dianion (Structure 17) which absorbed visible light
at 406 nm in the presence of nucleophiles (Figure 5).
Two absorbance readings were required to determine the
amount of compound 16 present in the crude material.
The first reading (A_{406}) measured the amount of free
dianion in the mixture. Hydroxide was then added to
completely hydrolyze compound 16 to its acid (Figure 5,
Structure 13) and the dianion, and a second reading was
taken (A_{406}(NaOH)). This determined the amount of
dianion liberated by the hydrolysis of this activated
ester of methoprene. The percent total ester in the
crude mixture was calculated using Equation 1.

$$\frac{A_{406}(NaOH) - A_{406}}{A_{406}(NaOH)} \times 100 = \text{Percent ester in mixture} \quad (1)$$

Figure 5 shows the UV spectra before and after
hydrolysis. This particular sample was composed of
about 60% of the active ester.
    In a similar manner the acylation of the protein
carrier with the HNSA-methoprene active ester was

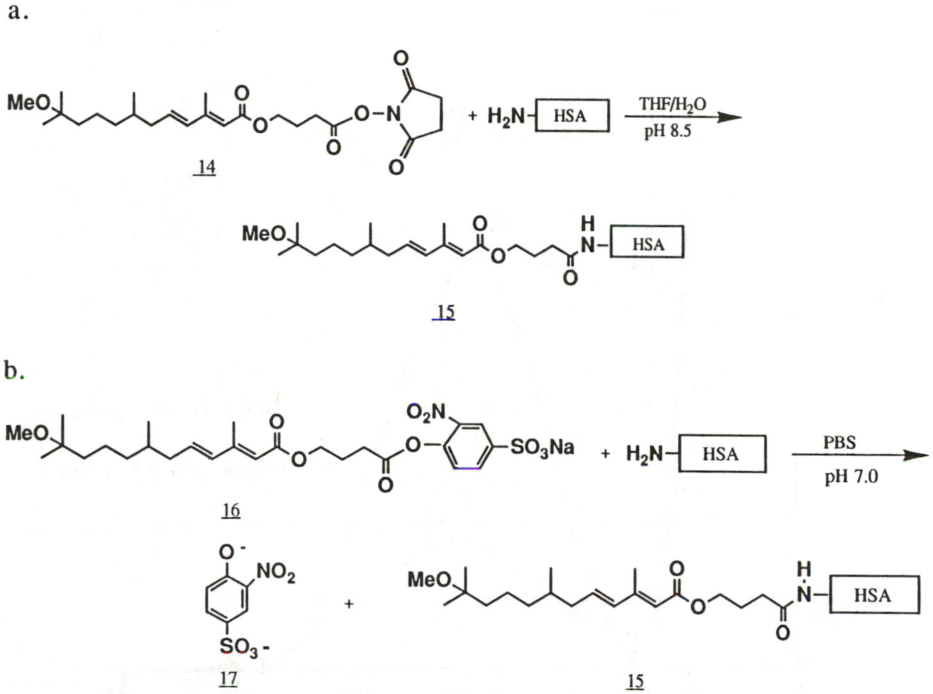

a.

b.

Figure 4. Preparation of the methoprene immunogen (Structure 15) by two methods: a. The NHS-ester of methoprene (Structure 14) was conjugated to human serum albumin (H₂N-HSA) in organic/aqueous solution. b. A water soluble active ester of methoprene (Structure 16) was prepared by the DCC coupling of methoprene-spacer acid (Structure 13) with 1-hydroxy-2-nitro-4-benzene sulfonate. Reaction of compound 16 with H₂N-HSA was carried out in aqueous phosphate buffered saline (PBS).

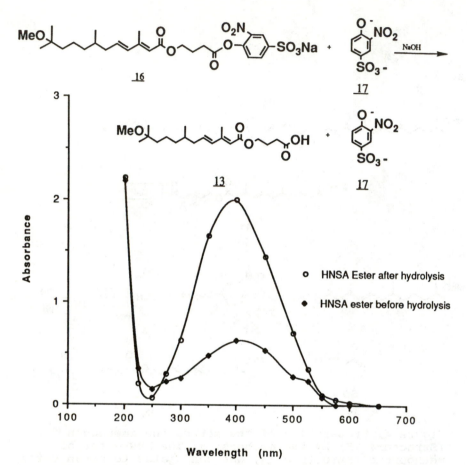

Figure 5. Hydrolysis reaction of HNSA-ester (Structure 16) with base gave methoprene-spacer acid (Structure 13) and the dianion (Structure 17). The spectrum shows the absorbance of the dianion before (♦) and after hydrolysis (o).

followed spectrophotometrically. The reaction of ester and protein was carried out in aqueous buffers at neutral pH (Figure 4b) due to the high degree of solubility of both reactants in aqueous solution. Generation of the free dianion was directly proportional to the amount of acylation of the protein. The calculation of hapten density was determined by measuring dianion increase over time. Using this method, and with greatly reduced reaction times, an average of 31 molecules of methoprene were introduced per molecule of protein (Table I). Hapten conjugates were dialyzed and lyophilized as described above.

Table I. Number of methoprene molecules
introduced per molecule of protein using
the HNSA activated ester

Reaction Time	Crude Ester		Protein	Dianion Liberated		Residues Reacted	
(min)	(mg)	(μmol)	(μmol)	$A_0{}^a$	$A_t{}^b$	$A_{406}{}^c$	$^{14}C^d$
15	10.52	21.86	0.072	0.750	0.837	34	31
30	8.11	16.86	0.064	0.579	0.640	32	28
180	8.50	17.66	0.068	0.607	0.656	27	28

a: Spectrophotometric absorbance reading taken at 406 nm to determine the amount of free dianion in the mixture at time = 0 min.
b: Second reading taken to determine the amount of dianion liberated by reaction of the active ester with protein at the time indicated.
c: The spectrophotometric method was used to determine hapten density.
d: $5-^{14}C$-methoprene was used to determine hapten density.

The two active ester methods for hapten conjugation are compared in Table II. The NHS-ester method relies on adjusting pH, solvent and time in order to obtain maximum hapten density. The HNSA-ester method is carried out in aqueous solution and is monitored easily to determine hapten density. Both activated ester methods produced suitable immunogens.

Table II. Comparison of hapten density by
different active ester methods

Method	Trial	Reaction Time	^{14}C	A_{406}
			Residues Reacted	
NHS-Ester	1	24 h	21	NA
	2	48 h	53	NA
	3	48 h	48	NA
HNSA-Ester	1	15 min	31	34
	2	30 min	28	32
	3	180 min	28	27

NA: Not applicable. NHS: N-hydroxysuccinimide. HNSA:
1-Hydroxy-2-nitro-4-benezene sulfonate. See Table I
for explanation of abbreviations.

ELISA Development

Female, New Zealand white rabbits were each immunized
with an emulsion of the methoprene immunogen (200 μg
per animal, 53 molecules of methoprene per molecule of
protein) in 250 μL phosphate buffered saline and 250 μL
Freund's complete adjuvant. A booster shot of the
immunogen (200 μg per animal) in Freund's incomplete
adjuvant was given to each animal one month after the
initial immunization and again two weeks after the
first booster shot. Rabbit anti-methoprene antiserum
was collected and used to develop a competition enzyme-
linked immunosorbent assay or cELISA for methoprene.
     The competition assay was designed which followed
the standard indirect ELISA format (17-18). The
methoprene conjugate was bound to a solid support in
the form of a microtiter plate. Free methoprene in
methanol (5 μL) was added to the pre-coated wells
followed by methoprene-specific antiserum. The
antibodies were allowed to compete for both immunogen-
bound and free methoprene. Enzyme-conjugated, goat-
antirabbit antibody was added, followed by substrate,
and the color was allowed to develop. The absorbance
of substrate over a range of methoprene concentrations
can be drawn as a standard curve, which is presented as
percent inhibition of the assay (Figure 6). The 50%
inhibition (I_{50}) of methoprene was at a concentration
of approximately 50 ng/mL.
     Cross reactivity of the methoprene antiserum was
tested using JH I and JH III, and methoprene

intermediates used to prepare the immunogen. JH I and JH III at concentrations up to 800 ng/mL did not cross react with the methoprene antiserum. By contrast, methoprene acid and methoprene-spacer acid were good competitors for the antiserum (Figure 7).

Application of the Methoprene cELISA

Methoprene residues are usually determined by standard analytical techniques such as HPLC (19-20) and GC (21-22). These techniques, however, remain costly, time consuming, and require extensive sample clean-up and preparation.

Extraction of Tobacco. Methoprene is used on tobacco against the cigarette beetle and the tobacco moth. Methoprene-treated tobacco samples were extracted following a procedure for the extraction of plant materials for determining methoprene residues by GC (21). Known amounts of methoprene in 1 mL methanol were added to 1 g portions of shredded tobacco, mixed well and allowed to thoroughly air dry. The spiked tobacco was then stirred with a 25 mL mixture of acetonitrile/water/Celite 45 (250 mL/30 mL/10 g). The mixture was filtered by suction and the filter cake was washed with acetonitrile/water. The filtrate was extracted with ether, distilled water, and sodium chloride. Ether extracts were combined and washed three times with distilled water, dried, filtered and the solvent removed. The residue was taken up into methanol (1 mL) and applied to the pre-coated microtiter plates (5 μL methanol/well), followed by the anti-methoprene antibody as described above.

Use of the cELISA to determine methoprene content of tobacco residues prepared as described above, showed that the extracts contained materials which interfered with the assay (Figure 8). Extracts of tobacco not only contained methoprene, but also plant substances that interfered with the assay. Thin-layer chromatography (TLC) was then used to clean up crude extracts from tobacco samples to reduce and/or eliminate interference.

Tobacco residues were resuspended in methanol (20 μL) and applied to TLC plates (2 μL/lane) along with a methoprene standard. The plates were developed with ethyl acetate:dichloromethane:hexane (1:2:7). The lane containing the standard was cut from the plate and developed in iodine. Regions containing methoprene (rf = 0.46) were scraped from the other lanes (approx. 1 cm^2/lane), and after adding hexane (400 μL) the scrapings were vortexed (1 min) and centrifuged (8000 x g, 15 min). The supernatants were removed to clean vials, evaporated by a stream of air, and resuspended

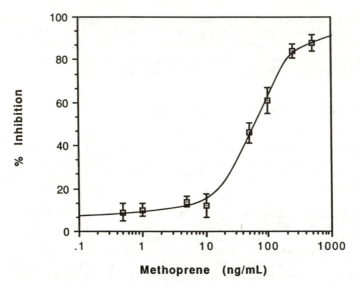

Figure 6. Percent inhibition of the antiserum by methoprene. 50% inhibition of the assay can be seen at about 50 ng/mL. Vertical bars indicate standard error (n=3 for each point).

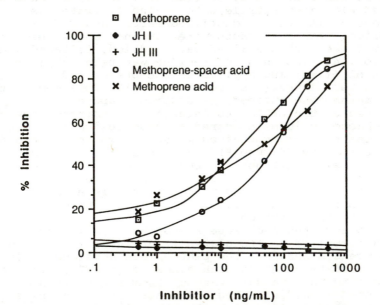

Figure 7. Cross reactivity of the methoprene antiserum with methoprene derivatives and juvenile hormones. The JHs showed virtually no cross reaction with the antiserum. Due to the complexity of the figure, standard error bars were omitted (n=4 for each point).

in methanol (1 mL). Aliquots (5 μL) of the purified
residue were applied to the methoprene cELISA.

Figure 9 shows the curve for tobacco residues
purified by TLC before analysis by the assay. It is
clear that the interference seen previously was
practically eliminated. However, such extensive sample
preparation makes use of this assay in its present form
cumbersome, at best. We are presently investigating an
alternative form of immunoassay, the enzyme immunoassay
(EIA) (6, 23). In this assay methoprene is conjugated
directly to an enzyme and the anti-methoprene antibody
is bound to the solid support. Free methoprene and
methoprene-enzyme conjugate are in solution and compete
for immobilized antibody binding sites. Unbound
methoprene is washed from the assay prior to addition
of substrate. Preliminary results under these
conditions indicate that tobacco extracts of
acetonitrile/water (9:1) do not require further
purification steps prior to application to the EIA.

Summary

The adaptation of immunochemical methods for many
classes of small compounds has made it possible to
develop assays for pesticide residue detection. A
cELISA was developed for methoprene, an analog of
insect juvenile hormone. Because of its size,
methoprene does not elicit an immune response by
itself. However, by conjugating methoprene to a
carrier protein it was made immunogenic in animals. A
four-carbon spacer group was incorporated between
methoprene and the carrier protein. The spacer was
first coupled to methoprene acid by a series of
protection/deprotection steps.

Two different activated ester methods were used to
conjugate methoprene to a protein. Both the NHS-ester
and the HNSA-ester methods produced immunogens of
suitable hapten density, however the HNSA-ester method
can be carried out in aqueous solution and allows for
the spectrophotometric determination of hapten density.
Rabbit polyclonal antiserum raised against the
methoprene immunogen was used in a cELISA to detect
free methoprene in the 5-300 ng/mL range. When
methoprene was extracted from treated tobacco samples,
the crude extracts contained materials which interfered
with the cELISA. Purification of tobacco residues by
TLC was found to sufficiently eliminate the
interference.

Acknowledgments

This research was funded, in part, by the R.J. Reynolds
Tobacco Company, contribution no. 2970 from the
Massachusetts Experiment Station.

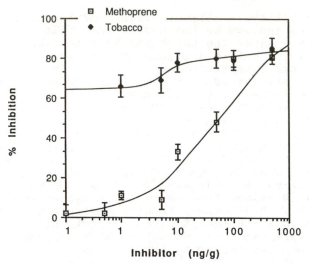

Figure 8. Methoprene treated-tobacco samples were extracted and the resulting residues were applied directly to the assay. The extracts contained materials which caused considerable interference with the assay. The competition of methoprene, in the absence of tobacco, was also plotted for comparison. Vertical bars indicate standard error (n=4 for each point).

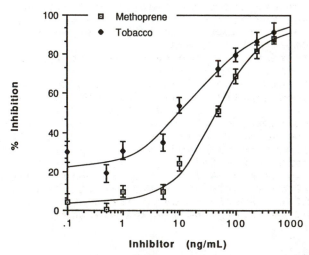

Figure 9. Purification of methoprene-treated tobacco residues by thin layer chromatography reduced the interference seen in Figure 8. Vertical bars indicate standard error (n=4 for each point).

Literature Cited

1. Williams, C.M. Sci. Am. 1967, 217, 13-17.
2. Bowers, W.S. Ent. Exp. Appl. 1982, 31, 3-14.
3. Staal,G.B. Ent. Exp. Appl. 1982, 31, 15-33.
4. Wakabayashi, N.; Waters, R.M. In Handbook of Natural Pesticides; Morgan, E.D.; Manduva, N.B. (Eds.); CRC Press: Boca Raton, FL, 1985; Vol. 3, pp. 87-151.
5. Herman, B. W. In Immunological Techniques in Insect Biology; Gilbert, L.I.; Miller, T.A. (Eds.); Springer-Verlag, NY, 1988; Chapter 5.
6. Nakamura, R.M., Voller A., Bidwell, D.E. In Immunochemistry; Weir, D.M., Herzenberg, L.A., Blackwell, C. (Eds.); Blackwell Scientific Publications, Oxford, 1986; Chapter 27.
7. Lauer, R.C.; Solomon, P.H.; Nakanishi, K; Erlanger, B.F. Experientia. 1974, 30, 558-560.
8. Baehr, J.C.; Pradelles, P.; Lebreux, C.; Cassier, P.; Dray, F. FEBS Lett. 1976, 69, 123-128.
9. Strambi, C.; Strambi, A; De Reggi, M.L.; Hirn, M.H.; Delaage, M.A. Eur. J. Biochem. 1981, 118, 401-406.
10. Kinsky, S.C.; Hashimoto, K.; Loader, J.E.; Benson, A.L. Biochim. Biophys. Acta. 1984, 769, 543-550.
11. Scott, D.; Nitecki, D.E.; Kindler, H.; Goodman, J.W. Mol. Immunol. 1984, 21, 1055-1060.
12. Sudo, R.; Kaneda, A.; Itoh, N. J. Org. Chem. 1967, 32, 1844-1846.
13. Sieber, P.; Andreatta, R.H.; Eisler, K.; Kamber, B.; Riniker, B.; Rink, H. In Peptides; Goodman, M.; Meienhofer, J., Eds.; Proceedings of the 5th Peptide Symposium: John Wiley and Sons, Inc.: New York, 1977; p. 543.
14. van Duzee, E.M.; Adkins, H. J. Am. Chem. Soc. 1935, 57, 147-151.
15. Mei, J.V. Masters of Science Thesis, University of Massachusetts, Amherst, MA, 1988.
16. Aldwin,L.; Nitecki, D.E. Anal. Biochem. 1987, 164, 494-501.
17. Voller, A.; Bidwell, D.E.; Bartlett, A. The Enzyme Linked Immunosorbent Assay (ELISA); Dynatech Laboratories, Inc.: Alexandria, VA, 1979; p.35.
18. DeToma, F.J.; MacDonald, B.A. Experimental Immunology: A Guidebook; MacMillan Publishing Co.: New York, 1987; p. 121.
19. Chamberlain, S.J. Analyst. 1985, 110, 879-880.
20. Heckman, R.A., Conner T.R. LC-GC. 1989, 7, 855-856.
21. Wright, J.E.; Jones, R.L. Bull. Environ. Contam. Toxicol. 1976, 15, 525-529.
22. Miller, W.W.; Wilkins, J.S.; Dunham, L.L. J. AOAC. 1975, 58, 10-14.
23. Kurstak, E. In Enzyme Immunodiagnosis; Academic Press: New York, 1986. Chapters 3 and 4.

RECEIVED April 16, 1990

Chapter 13

Barriers to Adopting Immunoassays in the Pesticide Analytical Laboratory

James N. Seiber[1], Qing Xiao Li[1], and Jeanette M. Van Emon[2]

[1]Department of Environmental Toxicology, University of California, Davis, CA 95616
[2]Environmental Monitoring Systems Laboratory, U.S. Environmental Protection Agency, Las Vegas, NV 89193-3478

Immunoassays offer much potential for rapid screening and quantitative analysis of pesticides in food and environmental samples. However, despite this potential, the field is still dominated by conventional analytical approaches based upon chromatographic and spectrometric methods. We examine some technical barriers to more widespread adoption and utilization of immunoassays, including method development time, amount of information delivered and inexplicable sources of error. Examples are provided for paraquat in relation to exposure assessment in farmworkers and food residue analyses; molinate in relation to low-level detection in surface waters; and bentazon in relation to specificity and sensitivity requirements built in to the immunizing antigen. A comparison of enzyme-linked immunosorbent assay (ELISA) results with those obtained from conventional methods will illustrate technical implementation barriers and suggest ways to overcome them.

A persisting problem in analyzing for trace levels of pesticides (and other toxicants) in human foods, animal tissues, and environmental samples lies in the cost and time involved in conducting the analysis. In the typical situation, samples are brought to the lab, preserved (usually by freezing), and then analyzed one-by-one over a period of time ranging from several days to several months. Frequently, the analytical costs are the major component of projects designed to assess exposures of people and wildlife, the environmental fate of chemicals in vegetation, air, water, or soil systems, the decontamination of toxic waste sites, etc. The costs include the time of technical personnel, reagent and solvent costs, and the capital cost of chromatographic and spectrometric instruments used in the final determination. Also, frequently the time involved in processing large sets of samples, at typical rates of 10-20 samples/working day, drives projects to timetables which provide results several months after the samples are taken. Ultra-low detection limits, while generally achievable by modifying conventional methods, may increase both the cost and time of the project.

Immunoassays (IAs) hold great potential for alleviating this situation in terms of decreasing the cost and time requirements and their contributing factors, and by providing results at low detection limits without major changes in procedure. Hammock and Mumma summarized the advantages of immunoassay(1):

0097–6156/90/0442–0156$06.00/0

- Detection limits - Picogram-nanogram levels are detectable in milliliter samples of incubation medium.
- Specificity - There is typically little cross-reactivity between the analyte and matrix, thus, minimizing the extent of cleanup needed.
- Speed - In addition to minimizing the time for sample preparation, the actual throughput for final determinations is on the order of 100 or more samples/hr [vs 5-10 for gas chromatography (GC) and high performance liquid chromatography (HPLC)].
- Cost effectiveness - With less time for sample preparation, less reagent cost per sample, and amenability to automation, the overall cost can be a fraction of that for a conventional analysis.
- Applicability - While applicable to virtually all classes of analytes, IAs work particularly well for those classes of polar and/or labile analytes which are not readily amenable to GC and HPLC analysis.

Indeed, the promise of IA has led to workable enzyme-linked immunosorbent assays (ELISAs) for a variety of pesticides, including thiolcarbamates, triazoles and triazines, substituted ureas and sulfonylureas, bipyridilium compounds, and other groups of chemicals(2) (Table I). Additionally, radioimmunoassay (RIA) methods exist for chlorinated hydrocarbons, organophosphates, phenoxy acids and other chemical classes.

However, the good points about IA are well described in other chapters in this book, and here we address the question: "Why do conventional GC and HPLC still dominate the field of trace analysis for pesticide residues, while IA has not yet made a major impact?" While remaining firm supporters of IA, we will play Devil's Advocate in addressing this subject, particularly in incorporating the viewpoint of the practicing residue chemist who has a variety of analytical techniques at his/her disposal.

The disadvantages of immunoassays fall into three broad categories:

1. Method Development Time - A new method based upon GC or HPLC can often be developed in less than one month - particularly for new chemicals belonging to a previously studied class or structural type. IA method developments, on the other hand, may require several months or several years as one synthesizes hapten derivatives, conjugates to protein, immunizes an animal to obtain antibodies, optimizes key assay parameters, and validates the final procedures. This becomes much less of a disadvantage if a stock of antibodies is available for the hapten of interest, in which case method development time for IA can be measured in weeks or a few months.

2. Amount of Information Delivered - IA is best suited to analysis of only one, or a few closely related analytes, in a sample while GC, HPLC, and particularly gas chromatography/mass spectrometry (GC/MS), can respond to dozens of analytes in a single run if properly calibrated.

Further, IA gives only a single piece of data (e.g. color or fluorescence intensity in ELISA) for each sample while GC, LC, or GC/MS gives us the analyte signal plus the signals from background scanned in the same chromatogram. This additional information provides important clues to the sample composition which can greatly aid in interpreting the result.

3. Inexplicable Pitfalls - Sometimes IA fails completely for reasons which are not easily identified. Conventional methods also fail, but the chromatogram may tell us immediately the cause of failure (e.g. flame not lit in an FID or FPD detector is signalled by a flat baseline; GC out of carrier gas leads to increase in RT; late-eluting peak overlaps chromatogram signalled by off-scale response, etc.). Of course, immunoassay failures can be diagnosed, too, by substituting fresh buffers, antibody preparations, plate lots, etc. and by running samples at several dilutions, but learning a whole new catalogue of trouble shooting symptoms represents

Table I. Representative Pesticides with Immunoassay

ELISA/EIA	RIA

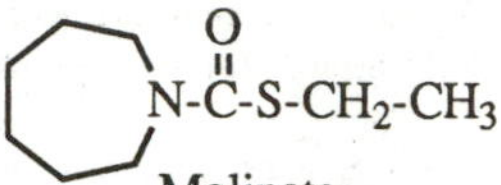

Molinate

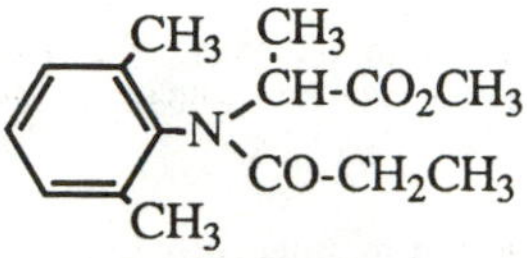

Metalaxyl

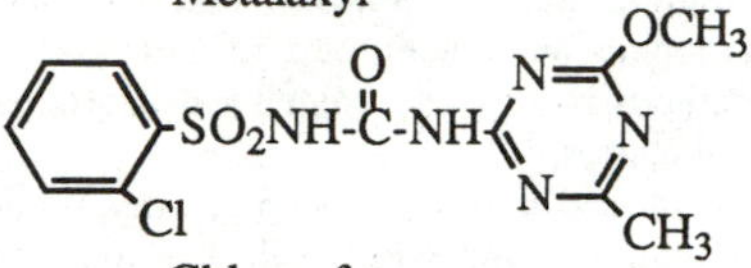

Chlorsufuron

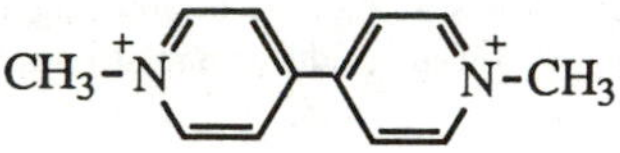

Paraquat

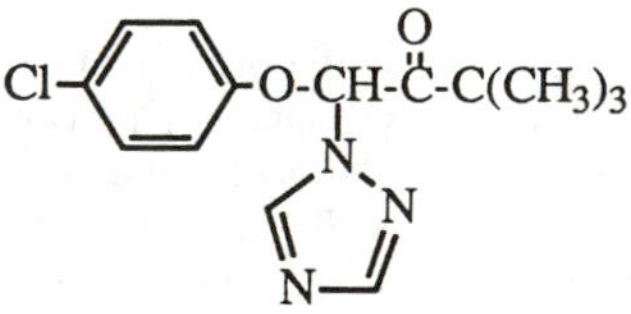

Triadimefon

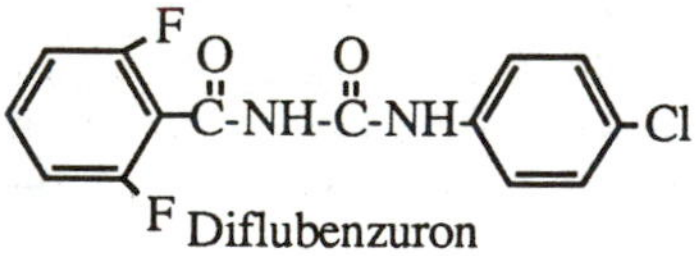

Diflubenzuron

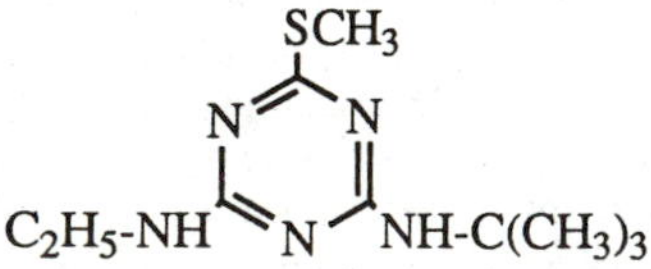

Terbutryn

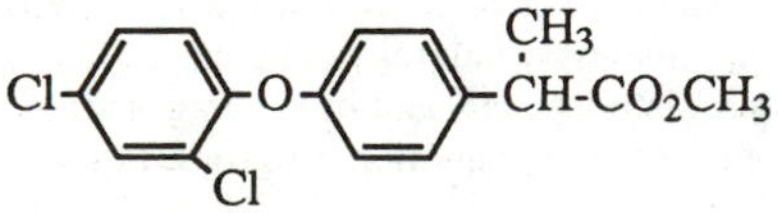

Diclofop Methyl

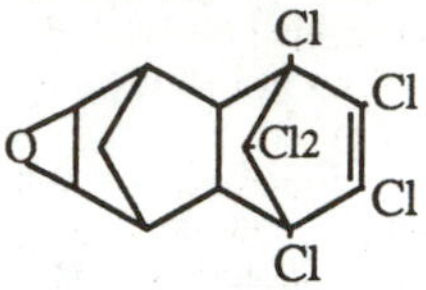

Dieldrin

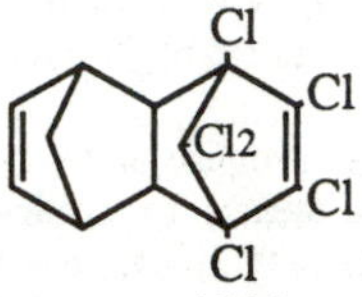

Aldrin

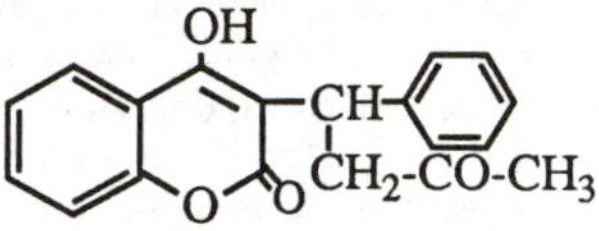

Warfarin

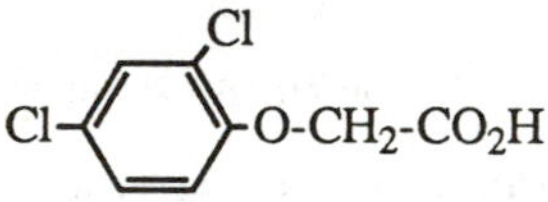

2,4-D

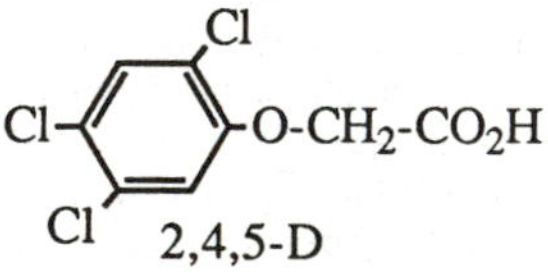

2,4,5-D

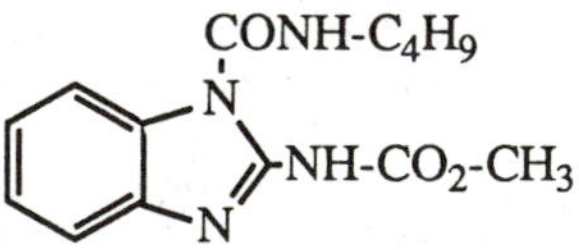

Benomyl

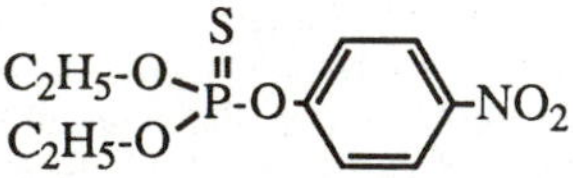

Parathion

a formidable barrier for the analytical laboratory which must produce results on a constrained timetable.

The following sections will provide examples of barriers to the adoption of immunoassay by practicing analysts along with some suggestions for reducing and overcoming them.

Paraquat. Paraquat is a widely used herbicide in cotton and potato culture, on orchard floors, and in landscape maintenance. There have been persistent reports of paraquat poisoning, mostly from ingestion both intentional and accidental, but also from dermal contact. There have also been reports linking paraquat to chronic intoxication, particularly in pulmonary disease from inhaling paraquat aerosols. In addition to food residue analysis in support of tolerance, worker exposure samples and ambient environmental samples are also of great interest.

Paraquat   I

Paraquat is an excellent candidate for immunoassay because it is ionic, and thus not amenable to conventional extraction and gas-liquid chromatography. HPLC analysis is also difficult because of the two separated charges, although ion-exchange HPLC and ion-pairing HPLC represent viable approaches to paraquat determination. In 1977, the first report of a successful IA was made using plasma as the matrix(3). Other reports of paraquat RIAs have appeared subsequently(4).

The standard method for analyzing paraquat is by ion-exchange cleanup followed by reduction to the radical cation, which is colored and read spectrophotometrically(5). The labor requirements and somewhat limited detectability of this method led us to develop a GC method based upon reduction with sodium borohydride to a mixture of the monoene and diene tertiary amines - both of which are amenable to GC on a deactivated column using an N-selective detector(6). The chromatogram shows both peaks (ratios may vary somewhat) which provides built-in confirmation of paraquat because of this dual response pattern. But the reduction-GC method is still a laborious one with a throughput of perhaps 8-10 samples/day under good conditions. It took nearly one man-year to develop and fully validate.

The paraquat ELISA developed in our laboratory started with the synthesis of the valeric acid derivative of paraquat (I) as hapten, and took well over one man-year to develop(7) (Figure 1). It is a very good method with fractional nanogram/mL sensitivity and (once fully validated) precision slightly better than the GC method. It is applicable to air filters, clothing patches, and hand washes, and showed promise for lymph and plasma analysis. The sample throughput was greater than GC and the method could be picked up rapidly by persons not skilled in the art. (Novices frequently pick up IA much faster than veteran analysts with a history of GC and HPLC experience!).

The barriers in this case were 3-fold:

1. Long Development Time - The ELISA method development exceeded 1 man-year, thus took even longer than for an unusually lengthy GC method development. Once again, had there been an available source of antibodies - a commercial equivalent to a fine chemical supplier or a chromatography supply house - ELISA development time would have been much reduced. Also, rather than having one individual master the synthetic chemistry, immunology, and analytical aspects of the development, using a teamwork approach combining the talents of several individuals could significantly reduce the 'real-time' requirement for IA development.

2. Assay Failure - Failures were experienced from time-to-time which were hard to diagnose. Generally, these were due to faulty reagents, buffers, and antibody solutions, or to defective plates.

Paraquat Sample
    | Incubate with
    ▼ Specific Ab
PQ_1-Ab Complex
Ab Excess
    | Incubate in PQ_2
    ▼ Sensitized cuvettes
PQ_1-Ab_1 Complex (free)
PQ_2-Ab_1 Complex (bound)
    ▼ Wash
PQ_2-Ab Complex (bound)
    | Incubate with Ab2-Enz
    ▼ (alkaline phosphatase)
PQ_2-Ab_1-Ab_2-Enz Complex
Ab_2-Enz Excess
    ▼ Wash
PQ_2-Ab_1-Ab_2-Enz Complex
    | Add substrate
    ▼ (p-nitrophenyl phosphate)
Yellow Color of p-Nitrophenol
    ▼ Read at 405nm
% Inhibition

Figure 1.Schematic of ELISA procedure for paraquat. (Reprinted from ref. 7. Copyright 1986 American Chemical Society.)

3. Regulatory Hurdles - Even though the method worked well for us, a regulatory agency probably would not have accepted the results because ELISA represents a non-standard methodology. This is still a real barrier, and drives many to 'conventional' methods even though they strongly suspect that IA might be even better.

Generally, paraquat represents a successful IA development, and a reasonably well accepted example of IA in trace pesticide analysis. Following the human exposure study, we provided additional applications to meat, milk, and potatoes(8). In each case, the ELISA provided detection limits lower than those available from spectroassay with minimal sample preparation (Table II).

In a recent statement of proposed research FDA singled out development and validation of ELISA for paraquat in potatoes as a target. This exemplifies the changing attitude toward IA by the regulatory community.

Table II. Detection Limits (ppm) for Paraquat in Foodstuffs by ELISA and Spectroassay

Sample	ELISA	Spectroassay
Milk	1 (0.5 g)[a]	10 (100-300 mL)[a]
Potato	0.8 (0.5 g)	10 (250 g)
Ground Beef	2.5 (0.5 g)	10 (50 g)

[a] Sample weight or volume required.
From Van Emon et al. (1987).

Molinate. The thiolcarbamate molinate also appeared to be a good target for ELISA but for different reasons than existed for paraquat. The herbicide is used extensively on rice in the Sacramento Valley and each year is detected as a contaminant in drainwaters from the field and in the Sacramento River. There are many water samples taken each year for molinate analysis and some other chemicals used for pest control on rice. Thus, an ELISA could be put to immediate use supplementing or perhaps replacing entirely the conventional method based upon extraction of water, followed by GC analysis of the extract.

$$\text{Molinate} \qquad \qquad \text{II}$$

Molinate: N-C(=O)-S-CH$_2$-CH$_3$ on azepane ring; II: N-C(=O)-S-CH$_2$CH$_2$COOH on azepane ring

The mercapto-propionic acid derivative (II) of molinate, conjugated to hemocyanin, provided the immunizing antigen for antibody production(9). The only significant cross-reactivity toward this antibody was exhibited by the mercapto-propionic acid derivative and some other closely related derivatives of molinate (Table III). Of these, only the sulfone is a metabolite of molinate in the environment. Thiobencarb, another thiolcarbamate rice herbicide, and still other thiolcarbamate showed essentially no cross-reactivity (Table III).

The GC procedure for molinate is quite simple and straightforward also, so that the method validation study focussed on comparing ELISA and GC(10). When four types of water were spiked with three concentrations of molinate, the water extracted by C-18 solid phase extraction (SPE) cartridge (neither GC or ELISA can detect molinate by direct analysis at less than about 20 ppb), then the SPE eluent analyzed by each of the two determination methods, recoveries were outstanding but GC consistently gave the better precision (Table IV). Most of the precisional error in ELISA was attributable to the coating antigen binding and to antigen-antibody reactions. However, even with these sources of imprecision, the variation of standard curves between ELISA runs was acceptable (Table V). Thus,

Table III. Inhibition of the Molinate ELISA by Some
Thiocarbamates and Related Compounds

Structure	% Cross-reactivity
(azepane ring)–N–C(=O)–SCH$_2$CH$_3$	100
(azepane ring)–N–C(=O)–SCH$_2$CH$_2$COOH	135
(azepane ring)–N–C(=O)–SCH$_2$(CH$_2$)$_3$CH$_2$COOH	318
(azepane ring)–N–C(=O)–SO$_2$–CH$_2$CH$_3$	15
(azepane ring)–N–C(=O)–SCH$_2$–C$_6$H$_4$–NH$_2$	28
(azepane ring)–N–C(=O)–SCH$_2$CH$_2$–C$_6$H$_4$–NH$_2$	35
(azepane ring)–NH	0
(diethyl)N–C(=O)–SCH$_2$–C$_6$H$_4$–Cl	0
(diisopropyl)N–C(=O)–SCH$_2$CH$_3$	0
(pyrrolidine)N–C(=O)–SCH$_2$CH$_3$	1.1
(C$_6$H$_5$)N(CH$_2$CH$_3$)–C(=O)–SCH$_2$CH$_3$	0.8
N(–CH$_2$CH$_3$)–C(=O)–SCH$_2$CH$_2$CH$_3$	1.4
(diethyl)N–C(=O)–SCH$_2$CH$_2$CH$_3$	1.0

another barrier to adoption of ELISA is lower precision when a really good GC method exists for the same analyte. Of course, this is really only a consideration when ELISA is the final quantitative assay, and not when ELISA is used for screening purposes - an end to which it is ideally suited.

We applied the two methods (GC and ELISA) to analyses of the organic extract of rice field soil, and here ELISA failed, giving roughly 2-3 times the value recorded and confirmed by a standard GC method. There was strong evidence for a matrix-derived interference producing the high positive proportionate error, but we have not yet figured out what the interference is or how to deal with it. It represents still another barrier, although one which could almost certainly be overcome by varying the soil extraction and/or cleanup procedures, that is by more method development.

We have already made the point that the chromatogram carries a much higher information load than the absorbance value - this is certainly true for molinate which chromatographs well and responds with high sensitivity to both N- and S- selective GC detectors and to GC-MS in the selective ion mode.

Table IV.   Recoveries of Molinate From Spiked Water Using Solid
Phase Extraction

Water	Conc.	%Recovery			
Type	ppm	GC		ELISA	
Tap	0.001	103.5	1.9	96.3	20.4
	0.010	99.7	2.1	91.7	4.6
Creek	0.010	96.0	3.7	94.0	10.6
Ditch	0.010	94.7	0.8	96.7	2.6
Field	1.000	96.9	3.2	111.0	7.2

Table V.   Standard Deviation of Molinate Standard Curves

Conc. (ng/mL)	3.9	7.8	15.6	31.2	62.5	125	250	500	1000	2000
Average* Inhibi-tion%	10.4	17.8	23.7	34.1	47.9	59.8	71.6	80.8	88.1	93.4
C.V.%	29.5	30.8	13.1	11.0	6.1	4.5	3.6	3.1	2.2	1.2

*The data were averages of nine standard curves.

Certainly, ELISA represents a viable screen for water samples because it is much faster than GC(11) and it is adaptable to field use. But even here, GC should be used to confirm ELISA positives (analyze a certain percentage of false positives only) picked up in a screening operation. On the other hand, in conducting research on the overall balance of molinate applied to a rice field agroecosystem (water, soil, air, vegetation), GC would be a better choice.

Bentazon. This cyclic sulfonamide is also used as a herbicide in rice fields to control, primarily, broad-leafed weeds. It is quite polar and also quite acidic at the sulfonamide NH. Bentazon illustrates the quandaries one faces in ELISA method development; which part of the molecule should support the protein bridge in the immunogen and which part should be left free to elicit antibody recognition. It is more convenient chemically to couple at the bentazon NH, but this is the area one would guess is unique and thus best left underivatized. We did couple at NH via the methylphenylacetic acid derivative, which was coupled to protein, activated it at the carboxy group, and reacted that with bentazon (at the NH again) to produce the antigen. That is, we followed two approaches to wind up with very similar (but not identical) antigens (Table VI). The synthetic chemistry possibilities are quite numerous - a major decision which is needed in producing candidate antigens.

Table VI. Approaches to Obtain Hapten-Protein Conjugate

I	II
1. Modify protein	1. Synthesize hapten
2. Activate modified protein	2. Activate hapten
3. Conjugate with target compound or its analogue	3. Conjugate with carrier protein

Bentazon assay optimization, which is quite lengthy, is currently underway in our laboratory, looking at such parameters as:

· Antibody concentration
· Coating antigen concentration
· Combination of coating-immunizing antigen
· pH
· Ionic strength
· Solvent tolerance
· Cross-reactivity

We have already found out something quite interesting - that N-methylbentazon is detected at 2 orders of magnitude lower detection limit than unmethylated bentazon (Figure 2) using antibodies produced from N-derivatized haptens. This is perhaps because the underivatized bentazon is ionized at the pH of the incubation medium and is not recognized by the antibodies originally produced by N-derivatized bentazon conjugates. Theory would predict the reverse will be true when coupling is done through the aromatic ring - that is, that bentazon would be better recognized by antibodies than N-methylbentazon. Preliminary experiments with antibodies induced by derivatives of 6- and 8-hydroxy bentazon have not provided substantiation for this supposition. So far, then, it appears that bentazon belongs to that poorly understood group of molecules which are poorly immunogenic - the ultimate barrier to developing an IA!

On the positive side, we presently have an ELISA for bentazon which requires methylation of the extract, same as is needed for the GC method. Thus, the ELISA for

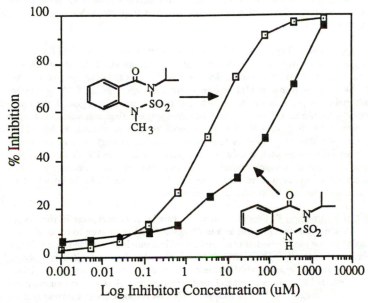

Figure 2. Standard Curves of Bentazon and Me-bentazon

bentazon is closer to the existing GC method, and there is a less cogent driving force for substituting ELISA for GC. When the extraction-cleanup-derivatization steps for ELISA are the same or similar to those for GC, ELISA becomes even less attractive to the 'conventional' analytical chemist.

CONCLUSION

The following is a summary of some of the major barriers toward adoption of IA by the analytical community, along with a statement, where appropriate, of an approach to mitigate or overcome the barrier.

Method Development Time. The approximately 1 man-year effort required for producing and selecting antibodies will remain a barrier for many firms and agencies which require an analytical method. Involvement of academic researchers, or the development of collaborative centers for immunoassays at Universities or at such agencies as EPA or USDA would certainly make this problem less acute. Also, as more experience is gained, the possibility of 'cookbook' procedures for synthesizing immunogen, eliciting antibodies and optimizing/ validating new IAs is quite real. Finally, antibody production should, in theory, be needed only once (providing the rabbit is reasonably prolific), through a network which makes antibody available for all such as exists for analytical standards in EPA's Pesticide Repository or through private sector supply houses. But even if antibody and procedure were distributed from a central source, method validation must still be done for IA in the lab where it is to be used, same as is needed for conventional methods.

Amount of Information Delivered. There are two barriers here, one in the single-analyte focus of IA which precludes its use as a multi-residue screen or even to cover all metabolites of a single toxicant, and the other is in the limited information contained in the final IA color or fluorescence intensity. For multi-residue screens, IAs have a limited role through such techniques as test batteries (several IAs in an integrated format) or antibody mixing. These are reasonable fields for future research and development. As far as information delivery is concerned, this limitation of IA will disappear when IA is coupled with conventional resolution techniques such as HPLC. The use of affinity columns and immobilized antibody-based detectors will produce a chromatogram with the peripheral vision analysts have come to expect from GC and HPLC.

Pitfalls. To a certain extent the pitfalls from unexpected failure of an IA method and from unanticipated matrix effects are simply creations of a too-high expectation for IA which perhaps has been aided by the enthusiasm of IA advocates. These 'barriers' will disappear when analysts recognize the obvious - that IA methods require running control and spiked samples same as conventional methods and, in fact, even more so. It goes practically without saying that IA does have one built-in quality control check that we assume all analysts employ - that is, the degree of agreement when two dilutions of the sample give the expected result (parallelism) from the standard curve. Also, sample replication will become more important for IA because of the somewhat ill-understood source of precisional error in IA. As always, imprecision is counterbalanced by increasing the number of replications - in the samples as well as the controls.

IA already has achieved a legitimate place in analytical labs as a screening tool and as a supplement to conventional methodologies such as multi-residue screens. More analytical labs will turn to IA when the above technical barriers are reduced, when antibody distribution becomes routine, and when agencies recognize IA as a legitimate analytical technique for regulatory purposes.

Literature Cited

1. Hammock, B.D. and Mumma, R.O. 1980. In: Recent Advances in Pesticide Analytical Methodology. J. Harvey, Jr. and G. Zweig, Eds., ACS Symposium Series 136, American Chemical Society, Washington, D.C., pp 321-352

2. Van Emon, J.M., Seiber, J.N., Hammock, B.D. 1989. In: Analytical Methods for Pesticides and Plant Growth Regulators. J. Sherma, Ed., New York, Academic Press, Vol.17, pp 217-263.

3. Levitt, T. 1977. Lancet 2(8033), 358.

4. Niewola, Z. 1985. Hayward, C., Symington, B.A., and Robson, R.T. Clin. Chim. Acta. 148:149-156.

5. Pack, D.E. 1967. In: Analytical Methods for Pesticides, Plant Growth Regulators, and Food Additives. Zweig, G., Ed., New York: Academic Press, Vol. 5, pp 473-481.

6. Seiber, J.N. and Woodrow, J.E. 1981. Arch. Environ. Contamin. Toxicol., 10:133-149.

7. Van Emon. J., Hammock, B., and Seiber J.N. 1986. Anal. Chem., 58:1866-1873.

8. Van Emon, J., Seiber, J.N., and Hammock, B. 1987. Bull. Environ. Contamin. Toxicol., 39:490-497.

9. Gee, S.J., Miyamoto, T., Goodrow, M.H., Buster, D., and Hammock, B.D. 1988. J. Agric. Food Chem., 36:863-870.

10. Li, Q.X., Gee, S.J., McChesney, M.M., Hammock, B.D., and Seiber, J.N. 1989. Anal. Chem., 61:819-823.

11. Stoddard, P. 1989. Development and Use of Immunochemical Assays By the California Department of Food and Agriculture. Presented at the 198th National Meeting of the American Chemical Society, AGRO 53, Sept. 10-15, 1989, Miami, FL.

RECEIVED June 5, 1990

IMMUNOASSAY ACTIVITIES IN INDUSTRY

Chapter 14

An Enzyme-Linked Immunosorbent Assay for Clomazone Herbicide

Ratna V. Dargar, John M. Tymonko, and Paul Van Der Werf[1]

Agricultural Chemical Group, FMC Corporation, Box 8, Princeton, NJ 08540

An enzyme-linked immunosorbent assay (ELISA) was developed for the quantitative analysis of clomazone, the active ingredient in Command herbicide. Antisera were obtained by immunizing rabbits with the bovine serum albumin conjugate of a p-amino analog of clomazone. Antibodies demonstrated specificity toward the clomazone moiety, with no cross-reactivity observed with any known soil metabolites of clomazone, or with 18 other soil applied pesticides that might be present in treated fields when clomazone is used. The detection limit for clomazone in soil was 10 ppb using a one-step aqueous extraction procedure. The ELISA analysis of soil samples spiked with clomazone correlated well with standard GLC methods, and with soil bioassay results (crop injury ratings) conducted under controlled greenhouse conditions. Under field conditions, the extent of crop injury was less predictive than desired, primarily due to a variety of hybrid, soil and environmental factors that impact the activity of this herbicide, and can reduce the predictive, field use of this ELISA procedure.

The use of immunoassays in the field of agricultural research has increased dramatically in recent years, and has become a reliable analytical tool that possesses numerous advantages over standard, chemical extraction and analytical methods. A few of these advantages (described in several review articles (1,2)), include its greater sensitivity and specificity, the increased speed of the assay, which allows greater sample through-put, the requirement for smaller samples for extraction, and the assay's improved cost effectiveness. Enzyme-linked immunosorbent assays (ELISA) have been

[1]Current address: I-Stat Corporation, 303 College Road East, Princeton, NJ 08540

developed for the analysis of several herbicides such as
chlorsulfuron (3), paraquat (4), diclofopmethyl (5) and atrazine
(6), and for the insecticides diflubenzuron (7) and paroxon (8).
      Clomazone [2-(2'-chlorophenyl)methyl-4,4-dimethyl-3-
isoxazolidinone] is the active ingredient in Command herbicide,
which is a soil applied herbicide produced by FMC Corporation, and
is used in soybeans to control many important grass and broadleaf
weeds (9). This herbicide is generally applied to the soil at rates
ranging from 0.75 to 1.25 lb a.i./a, depending upon the soil type at
the site, and the targeted weed species to be controlled. In
sensitive species, clomazone effectively inhibits the synthesis of
both chlorophyll and carotenoids, which results in injury symptoms
of white or chlorotic foliage in the plant. Due to soil residual
levels of clomazone, current label restrictions for Command
herbicide prohibit planting of cereal crops in the Fall of the year
that this herbicide is used, and also prohibits planting of seed
corn and cereals the following Spring, after clomazone applications.
Clomazone is generally safe to field corn when this crop is planted
9 months after herbicide application. However, occasional,
temporary injury has been observed on field corn in this time frame,
and has been related to factors such as differential corn hybrid
sensitivity (10), soil type, soil pH, rainfall levels, temperature
and herbicide misapplication. In an effort to understand the
various factors that influence clomazone persistence in the soil, a
rapid and highly specific immunoassay was developed, and will be
described here.

Test Methodology

Protein Conjugate Preparation.   Immunogen was prepared by the method
of Tijssen (11) with all procedures carried out at $0^{\circ}C$ with
stirring. One-half mM of the amino substituted analog of clomazone
(Figure 1) was dissolved in 15 mL 0.167 M hydrochloric acid with
0.052 M sodium nitrate added dropwise until a slight excess
developed as indicated by a starch-iodide test. After 30 minutes,
5.0 mL of the diazotized analog of the clomazone mixture was slowly
added to bovine serum albumin (BSA (25 mL, 5 mg/mL in 0.1 M sodium
borate, pH 9.0)) with the pH maintained by addition of 5N NaOH.
After 2 hours the mixture was dialyzed against 10 mM sodium
phosphate (pH 7.4) containing 0.02% sodium azide. The substitution
ratio of the clomazone analog to BSA was calculated to be 19.2
using the equations of Fenton and Singer (12) with the assumption of
no loss of protein. A rabbit serum albumin (RSA) conjugate was
prepared by the same procedure, except the concentration of RSA used
in the coupling step was 20 mg/mL, and yielded a substitution ratio
of 2.4.

Antiserum Preparation.   Anti-clomazone antisera were prepared by
Pel-Freeze Biologicals, Rogers, AR. Three New Zealand white rabbits
were inoculated in several intradermal sites with a total of 1 mg
BSA conjugate (1:1 emulsification in Freund's complete adjuvant) at
0, 2 and 4 weeks. Boosters (1 mg intradermal and 0.5 mg
intravenous) were given at 4 to 6 week intervals, with the animals
bled 10 to 14 days later.

ELISA Procedure. All incubations were carried out at ambient
temperature in a chamber containing moistened paper towels.
Clomazone standards were prepared by diluting a stock solution
(1 mg/mL in Tris-buffered saline (TBS)) with deionized water to
final concentrations ranging from 1 to 250 ng/mL. To immobilize the
hapten, all microtiter wells (Dynatech 96 'U' polyvinyl chloride
plates) except 3, were incubated overnight with 200 uL of TBS (TBS,
20 mM Tris:HCl, 146 mM NaCl, pH 7.4) containing 0.3 ug/mL RSA
conjugate. The remaining 3 wells (blanks) contained 200 uL of TBS.
The wells were washed 3 times by using a vacuum aspirator to fill
and empty the wells with washing buffer (TBS containing 0.05% Tween
20, adjuvant). The wells were then filled with TBS containing 0.1%
gelatin and 0.05% sodium azide for 1 to 1.5 hours to block the
unoccupied protein-binding sites, and then washed with washing
buffer as above. Unknown samples or standard solutions (100 uL)
followed by 100 uL of antiserum solution (prepared by diluting anti-
clomazone antiserum 1/40,0000 in 2X washing buffer containing 0.1%
gelatin) were added to the wells. Maximum sensitivity of the assay
was obtained by decreasing the antigen concentration, as indicated
by Hassan, et. al.(13) to ensure that the antibody levels were low
enough to become rate limiting. Following a 1 hour incubation, the
wells were washed as above. Goat anti-rabbit IgG-alkaline
phosphatase (IgG-AP) (200 uL) was diluted 1/1000 in washing buffer
containing 0.1% gelatin and added to the wells 2 hours after they
were washed. Substrate (215 uL, 1 mg/mL p-nitrophenyl phosphate in
1 M diethanolamine, 1 mM MgCl2, pH 9.8) was then added, and after 30
minutes 30 uL 5 N NaOH was also added to stop the reaction.
Absorbances of unknowns and a set of standards on each plate were
measured on a Dynatech MR650 reader (410 nm filter) and
concentrations of clomazone calculated automatically using
Immunosoft software, based on the concentration of the standards.
During the validation of the test, data was calculated manually and
was reported as percent inhibition of color development.

Soil Extraction. Approximately 1 gram of soil and 4 mL of distilled
water were thoroughly mixed in polypropylene tubes (17 X 100 mm) by
vortex suspension, allowed to sit overnight, mixed again, and were
then centrifuged at 10,000 rpm for 15 minutes. The supernatants
were used in the ELISA assay with no further preparation. For GLC
analysis, 1 mL aliquotes of the aqueous phase were extracted with
four 1 mL portions of hexane which were combined, concentrated under
nitrogen and adjusted to a 1 mL volume with hexane. Analyses of
hexane extracts were made on a Hewlett-Packard GLC with a DB-5
column (0.530 mm X 15mm).

Greenhouse Bioassay. Sandy loam soil was treated with clomazone at
rates ranging from 0.063 to 1.0 kg/ha and thoroughly mixed to
incorporate the chemical into the soil. The bioassay was conducted
by planting wheat (Triticum aestivum) and velvetleaf (Abutilon
theophrasti) into the test soils and visually assessing plant injury
in the form of bleaching (0 to 100% scale) 2 weeks after planting.

Results and Discussion

Assay Development and Validation. Reproducibility of this ELISA
assay was determined based on a set of clomazone standards that were
run on different plates on the same day (intra-assay) and on
different days (inter-assay). The intra-assay coefficient of
variation of the standards changed from 1.5% at the highest
clomazone concentration (250 ppb) to 22% at the lowest concentration
of 1.4 ppb. The coefficient of variation(CV) at clomazone rate of
12.3 ppb was 10%. Similar values were obtained for the inter-assay
variability, with the CV of the 1.4 ppb concentration being 22.3%,
and the CV for the 250 ppb concentration being 2.7%. The CV for the
10 ppb concentration of clomazone was about 5% between tests.
Analysis of the data for this range of clomazone concentrations
indicates that there is good correlation (r^2=0.97) between the log
of the concentration of clomazone and percent inhibition in the
assay when the linear regression equation was used. Based on these
results, the limit of the test's sensitivity was defined as 2 ppb
(10 ppb in soil) and the limit of detection was set at 1 ppb.

Specificity of this ELISA assay was explored in detail, with
various analogs tested that possessed chemical substitutions on both
the phenyl ring, and on the isoxazolidinone ring (14). As indicated
in the paper by Dargar et. al.(14), little, or no cross reactivity
was observed with compounds substituted in the isoxazolidinone ring,
while some cross reactivity was observed with analogs substituted in
the phenyl ring. Cross reactivity from the phenyl substituted
analogs should be expected, as the amino analog was used as the
haptan, to ensure greater specificity toward the unique
isoxazolidinone ring of clomazone. No cross reactivity was found
with the only known soil metabolite of clomazone (15), which is
shown in Figure 1, at rates up to 2,000 ppb, nor to 18 commercial
herbicides that could be expected to be present in agricultural
fields where clomazone is in use (14).

Most of the developmental work involved in this assay was
carried out using a technical grade of clomazone. As this herbicide
is sold commercially as a 4 lb/gallon emulsifiable concentrate (EC),
the technical and EC formulations were compared in the assay to
ensure that the formulated product did not adversely affect the
ELISA results. When similar concentrations of both formulations
were tested and compared by linear regression analysis, the data
indicated that there was no significant difference between these
formulations, with an r^2 value for this analysis of 0.99.

Soil samples that were spiked with clomazone at rates of 0.5 to
5.0 ug/g of soil were extracted with water, and then analyzed
directly by ELISA, or, the aqueous extracts were further extracted
with hexane for analysis by GLC. Within the limits of detection set
for this assay, levels of clomazone recovered from the soil with
water was excellent, except at very low soil residual levels of
clomazone as shown in Figure 2. The linear regression analysis for
this recovery experiment is described by the equation "Equation 1",
which has an r^2 value of 0.99.

$$Y = 1.094X - 76.28 \tag{1}$$

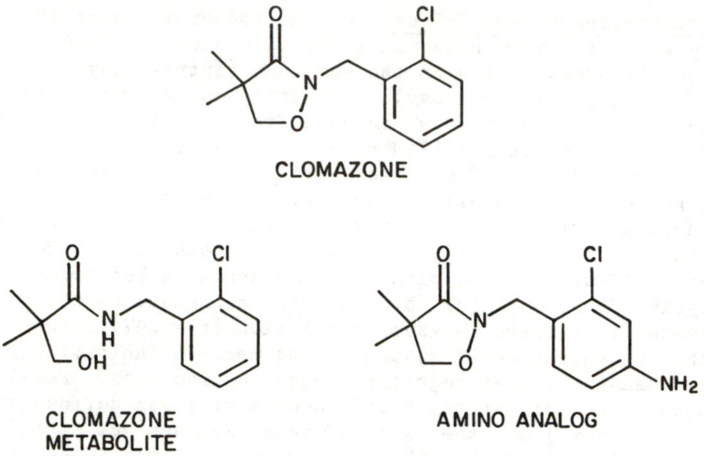

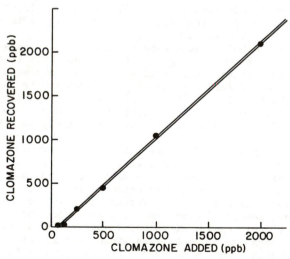

CLOMAZONE

CLOMAZONE
METABOLITE

AMINO ANALOG

Figure 1.   Structures of clomazone, the amino analog and the
soil metabolite of clomazone.

Figure 2.   Soil recovery of clomazone from sandy loam soil.

The linear regression analysis comparing results from a separate experiment designed to compare this ELISA and the standard GLC methodology indicate that the ELISA assay and GLC method are highly correlated (Figure 3), and described by the linear model in "Equation 2" which has an r^2 value for the equation of 0.98.

$$Y = 1.29X + 76.12 \qquad (2)$$

Clomazone Concentrations and Bioassay Results. One of the major objectives in the development of this ELISA assay was to obtain a rapid method of defining soil residual levels of clomazone under field situations, and use this data as a predictive tool in identifying fields that possessed greater potential for causing injury to sensitive rotational crops. This objective would require that the results of the soil analysis correlate with expected results (crop injury) that are observed under field conditions. An intermediate step toward this objective was completed by comparing the ELISA results from soils spiked with various concentrations of clomazone, with bioassay results that were completed in the greenhouse. Under these controlled environmental conditions, a plot of the logarithm of the concentration of clomazone from the ELISA results versus the percent injury of wheat was made (Figure 4). The plot shows a typical sigmoidal dose response curve, which indicates that there is a good correlation between the ELISA assay and the bioassay results that were obtained from the same soil samples.

Although there is good correlation of the ELISA results with both the standard GLC methods and with the bioassay results conducted under controlled conditions, unpublished results with clomazone indicate that the correlation of any soil analytical results (ELISA or GLC) with actual crop injury that is observed in the field is poor. This poor correlation between observed soil levels of clomazone, and the actual occurance of carryover injury in the field, indicates that factors unrelated to the soil analytical methods used, have a large impact on the ability to use soil residual levels as a predictive tool for this herbicide. An example of the impact of other factors can be seen with corn, where unpublished results attribute the poor correlation between detected soil levels of clomazone and observed injury in the field to several factors, which include differences in corn sensitivity toward clomazone, the method of clomazone application, application rate, soil type/organic matter levels, soil pH, soil sampling methods, time of corn injury assessment, interactions with other pesticides and environmental factors.

An example of the impact of just one of these factors on the correlation of field injury to soil analytical results is shown in Figure 5, which demonstrates the impact of corn hybrid sensitivity toward clomazone. This data indicates that corn hybrid sensitivity varies greatly when hybrids are tested under uniform conditions in the greenhouse. The figure demonstrates that there is a range in sensitivity toward clomazone that spans a ten-fold range in application rates between the most tolerant and more sensitive corn hybrids that were tested (10). This extreme range in sensitivity of corn toward clomazone can result in large differences in observed injury in a field with a known level of clomazone, with some corn

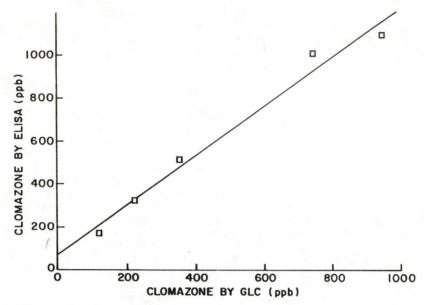

Figure 3.   Comparison of results obtained from the standard GLC technique used to define soil levels of clomazone, and results from ELISA.

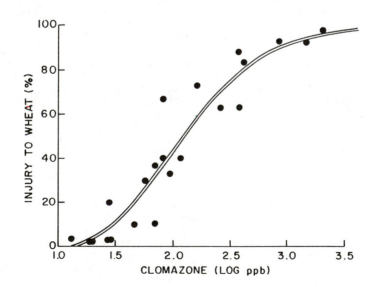

Figure 4.   Plot of clomazone residual soil levels determined by ELISA vs wheat injury from greenhouse bioassay results.

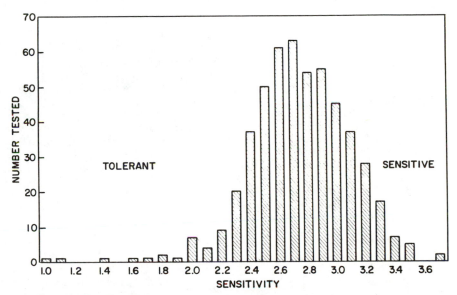

Figure 5. Distribution of corn hybrid sensitivity toward clomazone. Sensitivity = -Ln (Rate (Kg/Ha) to cause 20% Discoloration).

hybrids showing no injury symptoms caused by clomazone, while other, more sensitive hybrids might be injured. Thus, this one factor (hybrid sensitivity) can drastically effect the potential of any soil analytical technique to be used as a predictive tool to define when rotational crop injury might occur in the field. If the other factors that were listed above are also taken into account, the correlation between soil levels of clomazone and field injury levels can become even less precise.

Conclusions

The results have shown that we have developed a sensitive and specific ELISA assay for the analysis of clomazone residues from soil samples. The procedures have demonstrated good recovery of clomazone from soil, and excellent correlation of the ELISA test results with standard GLC methodology. In addition, the results of the ELISA tests demonstrate good correlation between the observed soil levels of clomazone, and crop injury when the bioassay is performed under controlled, greenhouse conditions. This assay could, therefore, be used as a more rapid and convenient analytical method over the standard GLC technique after further validation. The impact of factors not related to soil analytical methods, however, limits the convenience and use of this ELISA assay for it's main initial objective, which was to serve as a predictive tool to identify sites with potential for causing injury to rotational crops following clomazone applications.

Acknowledgments

Acknowledgments are provided to Dr. David W. Keifer for his information about corn hybrid sensitivity toward clomazone, and factors that affect their sensitivity.

Literature Cited

1. Hammock, R. R.; Mumma, R. O. Pesticide Analytical Methodology; American Chemical Society: Washington, DC, 1980; p 321.
2. Harrison, R. O.; Gee, S. J.; Hammock, B. D. In Biotechnology for Crop Protection; Hedin, P. A. et. al., Ed.; ACS Symposium Series No. 379; American Chemical Society: Washington, DC, 1987; p 316.
3. Kelley, M. M.; Zahnow, E. W.; Petersen, W. C.; Toy, S. T. J. Agric. Food Chem. 1985, 33, 962-965.
4. Van Emon, J.; Hammock, B.; Seiber, J. N. Anal. Chem. 1986, 58, 1866-1873.
5. Schwalbe, M.; Dorn, E.; Beyermann, K. J. Agric. Food Chem. 1984, 32, 734-741.
6. Huber, S. J. Chemosphere 1985, 14, 1795-1803.
7. Wie, S. I.; Hammock, B. D. J. Agric. Food Chem. 1982, 30, 949-957.
8. Brimfield, A. A.; Lenz, D. E.; Graham, C.; Hunter, K. W. J. Agric. Food Chem. 1985, 33, 1237-1242.
9. Keifer, D. W. North Central Weed Control Conf. Proc., 1988.

10. Chang, J. H.; Konz, M. J. ACS Abstracts, 187th National Meeting 1983, No. 22.
11. Tijssen, P. Practice and Theory of Enzyme Immunoassays; Elsevier Science Publishers; Amsterdam, The Netherlands, 1985, p 279.
12. Fenton, J. W. II; Singer, S. J. Biochemistry 1971, 10, 14291-14307.
13. Hassan, F.; N. E. Rothnie; S. P. Yeung; M. V. Palmer. J. Agric. Food Chem. 1988, 36, 398-403.
14. Dargar, R. V.; J. M. Tymonko; P. VanDerWerf. J. Agric Food Chem. In press.
15. Froelich, L. W.; T. A. Bixler; R. A. Robinson. North Central Weed Control Conf. Proc., 1984. Vol 39:79.

RECEIVED June 5, 1990

Chapter 15

Immunoassay Detection Methods for Alachlor
Application to Analysis of Environmental Water Samples

Paul C. C. Feng, Stephen J. Wratten, Eugene W. Logusch,
Susan R. Horton, and C. Ray Sharp

Monsanto Agricultural Company, St. Louis, MO 63198

Alachlor was covalently attached to various proteins by a procedure involving protein thiolation, followed by nucleophilic displacement at the chloroacetamide moiety. An alachlor conjugate of sheep γ-immunoglobulin (IgG) was used to induce the production in rabbits of polyclonal antibodies directed toward alachlor. Cross-reactivity studies showed that these antibodies were capable of distinguishing alachlor from other structurally similar chloroacetanilide herbicides, including metolachlor, and that they showed little cross-reactivity toward the major soil metabolites of alachlor. The antibodies were used to develop an enzyme-linked immunosorbent assay (ELISA) for alachlor in water with a detection range of 0.2 to 8.0 ppb. The concentration of alachlor in environmental water samples was measured using the ELISA and the results were compared with those obtained by gas chromatography/mass spectrometry (GC/MS) methods.

The development of immunoassay detection methods for pesticides and other environmental chemicals is a rapidly growing area of investigation. Immunoassays offer certain advantages over conventional instrumental methods for the analysis of pesticide residues (1-5). Immunoassays are often highly efficient and cost-effective, and are ideally suited for screening large numbers of samples for low levels of specific analytes. The detection of pesticides in drinking water provides an example of residue analysis in which immunoassay methods might compare favorably with other analytical techniques.

Alachlor (Figure 1) is the active ingredient of Lasso herbicide and several other herbicide products, and is one of the most widely used of the chloroacetanilide herbicides. We have recently developed effective methods for covalent conjugation of alachlor to protein carriers (6), and will describe the application of this methodology to the development of antibodies to alachlor. We report herein on the development of an inhibition ELISA technique for detecting

0097–6156/90/0442–0180$06.00/0

alachlor which is based on polyclonal rabbit antibodies raised against an alachlor conjugate prepared with sheep γ-immunoglobulin. The utility of the immunoassay method is illustrated for the detection of alachlor in environmental water samples, and is compared with an established GC/MS method. Our results indicate that the alachlor ELISA can be effectively utilized as a primary screen to select environmental samples for confirmatory instrumental verification of the presence of alachlor at low ppb levels.

Materials and Methods

Materials. Bovine serum albumin (BSA), sheep γ-immunoglobulin (IgG), and o-phenylene diamine (PDA) were purchased from Sigma Chemical Co. S-Acetylmercaptosuccinic anhydride (AMSA) and N-acetylhomocysteine thiolactone (AHT) were obtained from Aldrich Chemical Co. Immulon-1 96-well flat-bottom microtitre plates were purchased from Dynatech. Goat anti-rabbit γ-globulin conjugated to horseradish peroxide (GAR-HRP) was obtained from Cooper Biomedical Co. Freund's complete and incomplete adjuvants were obtained from Difco Laboratories. Non-fat dry milk powder (Food Club brand) was obtained locally. Uniformly phenyl-^{14}C-labeled alachlor was obtained from New England Nuclear, and showed greater than 98% radiological and chemical purity; it was diluted to a specific activitv of 0.06 mCi/mmol prior to use.

Instrumentation. Spectrophotometric absorbances of 96 well microtitre plates were recorded using a Bio-Tek EI 310 reader equipped with a 490 nm filter. The plates were washed using a Dynatech Dynawasher II. A 12-channel Titertek pipet (50 to 200 μL) from Flow Laboratories was used for dispensing liquids.

GC/MS Analysis of Alachlor. Quantitation of alachlor in water samples was carried out using an established GC/MS method. Deuterated alachlor, used as an internal standard, was spiked into 1.0 L samples of water. Organic soluble materials in the samples were extracted using a reverse-phase (C-18) solid-phase extraction filter (J. T. Baker Co.) and were afterwards eluted with a solvent mixture (1.0 mL) consisting of 5:45:50 ethyl acetate:isooctane:methylene chloride. Instrumental analyses were performed on a Finnigan model 4535 quadrupole GC/MS in the electron impact mode. Samples entered the mass spectrometer after injection onto a 15 m DB-5 capillary column (90 °C for 1 min, 90-120 °C at 10 °C/min, 120-140 °C at 2 °C/min). Alachlor was detected by selected ion monitoring of two characteristic fragment ions (m/z 160 and 188), and quantitated by comparison with the corresponding deuterated ions (m/z 171 and 199) arising from the internal standard. The sensitivity of this assay was established at 0.2 ppb of alachlor contained in the raw water sample.

Synthesis of Alachlor Protein Conjugates. Radiolabeled alachlor (hapten) was covalently attached to thiolated BSA and sheep IgG. Sulfhydryl groups were introduced onto lysine residues of BSA using the thiolating agent AHT, and onto the lysine residues of IgG using the thiolating agent AMSA. The protein (200 mg of BSA or IgG) and 25 molar equivalents of AHT or AMSA were dissolved in water (6 mL) at 0 °C, to which alachlor (25 equivalents) dissolved in dioxane (1 mL) was slowly added. Sodium carbonate-bicarbonate buffer (1.0 M, pH 11) was then added to adjust the pH to 11 and the reaction mixture was stirred at 0 °C for 15 min. After 2 hours of additional stirring at 50 °C, the reaction mixture was neutralized and the alachlor-protein conjugate was purified by a 24 hour dialysis against running water or by Sephadex G-25 size exclusion chromatography (2 x 50 cm column using 0.2 M NaCl). Both methods effectively separated the alachlor-protein conjugates from excess alachlor and thiolating agents. The radioactivity of alachlor-protein conjugates was determined by liquid scintillation counting. The protein concentrations of BSA and IgG were calculated from their UV absorbance at 280 nm and their molar extinction coefficients (39 and 188 mM/cm for BSA and IgG, respectively). Based on radioactivity analysis, 12 and 19 moles of alachlor were conjugated per mole of BSA and IgG, respectively. The IgG conjugate was used as the immunizing antigen in rabbits and the BSA conjugate was used as the coating antigen in ELISA. The alachlor-protein conjugates were lyophilized and stored at -20 °C.

Antibody Generation. The IgG conjugate of alachlor (1.0 mg) was dissolved in 0.3 mL of phosphate buffered saline (PBS; pH 7.4, 0.01 M phosphate and 0.15 M NaCl). The resulting solution was emulsified with Freund's complete adjuvant (1.0 mL), and was then injected intradermally into female New Zealand white rabbits. The animals were immunized with 1.0 mg of the immunogen and boosted at 4-6 week intervals with 0.1 to 0.5 mg of the same immunogen in Freund's incomplete adjuvant. Whole blood (25 mL) was obtained 2 weeks after each boost by bleeding from the ear vein, allowed to coagulate overnight at 4 °C, and centrifuged to generate the serum. Aliquots of the sera were stored at -80 °C.

Immunoassays. A checkerboard assay (7) was initially conducted with sera obtained from different bleeds and different animals to select the serum with the highest titre of antibodies. The checkerboard assay consisted of reacting varied amounts of the coating antigen (alachlor-BSA) with varied concentrations of the antiserum to establish the most sensitive combination of serum and coating antigen concentrations to be used in the inhibition ELISA. For alachlor, this concentration was established to be 5 ng/well of the coating antigen and 3,500 fold dilution of the selected serum. Plates coated with the coating antigen were stored desiccated at -20 °C, and remained stable for at least 4 months.

Inhibition ELISA was conducted in the following manner. A solution of the coating antigen, consisting of 5 ng of alachlor-BSA in 0.1 mL of sodium carbonate-bicarbonate buffer (0.05 M, pH 9.6), was dispensed into each well of 96-well microtitre plates and stored for 12 hours at 4 °C. The unbound coating antigen was then removed from the wells by washing the plates three times with PBS. The plates could be stored desiccated at -20 °C at this point. For ELISA preparation, the remaining active sites in the wells were blocked with a solution of 8.0% powdered milk in PBS (0.3 mL) for 1 hour at 22 °C.

Serum previously stored at -80 °C was freshly thawed and diluted 3,500 fold using PBS-T (0.02% Tween 20 in PBS). An aqueous alachlor standard or unknown sample was preincubated with an equal volume of diluted serum at 22 °C for 1 hour. This mixture (0.1 mL/well) was then dispensed into six replicate wells on the plate, which was then covered and incubated at 22 °C for 1.5 hours. After triplicate washes of the wells with PBS-T, each well was treated with 0.1 mL of GAR-HRP (freshly thawed and diluted 4,000 fold with 1.0% powdered milk in PBS). After a final cycle of four washes with PBS-T, 0.2 mL of freshly prepared PDA substrate solution (0.4 mg/mL PDA and 0.01% H_2O_2 in 0.05 M citric acid containing 0.15 M sodium dibasic phosphate, pH 5.0) was dispensed into each of the wells and incubated in the dark at 22 °C for 30-60 minutes. Sulfuric acid (4 N, 0.05 mL/well) was added to stop the reaction, and the final absorbance of each well (490 nm) was recorded. The presence of free alachlor in samples inhibited the binding of the antibody to alachlor-BSA, resulting in an inhibition of the development of absorbance at 490 nm. The amount of free alachlor was thus inversely proportional to the intensity of color produced. The level of alachlor in unknown water samples was calculated based on alachlor standards which had been analyzed simultaneously on each plate. Alachlor standards (0, 0.2, 0.5, 1.0, 3.0, 5.0, and 8.0 ppb in deionized water) were stored at -20 °C in 1 mL portions, and were freshly thawed for each assay.

Six wells on each plate not coated with the coating antigen served as background wells. Seven alachlor standards and eight samples were each analyzed in six replicate wells per plate. The absorbances of the replicate wells measured by the Bio-Tek reader were used in the following calculations. The median was calculated for the six replicate background wells and was then subtracted from the medians of the standards and samples. All the resulting median values of standards and samples were divided by the median of the standard without alachlor (0 ppb) to generate percentages of absorbance. The percentages of absorbance of the alachlor standards were plotted against the ppb concentrations of alachlor in the sample on a logarithmic scale. A hyperbolic curve was fitted to the data to generate the standard curve, which was then used to calculate concentrations of alachlor in unknown samples.

Cross-Reactivity Studies. The reactivity of the antibodies with a series of alachlor analogues was examined. The concentration of an analyte producing a 50% inhibition in absorbance in the ELISA was defined as its IC_{50} value (50% inhibition concentration). The IC_{50} value of alachlor in picomoles per mL was divided by the corresponding value from the analyte and multiplied by 100 to produce the percentage cross-reactivity values. The percentage cross-reactivity of the antibodies to alachlor calculated in this way was 100%.

Results and Discussion

Generation of Antibodies. Because of their small molecular weight, pesticides such as alachlor are not generally immunogenic ([4]). A key step in immunoassay development therefore involves the covalent conjugation of the pesticide or an appropriate analogue to a carrier protein ([6]). In considering various approaches to alachlor hapten-protein conjugation, we thought it desirable to utilize for attachment the single functional group common to all chloro-acetamide herbicides, i.e. the chloroacetamide moiety. Such an approach would in principle leave free the aromatic ring and methoxymethyl side chain, thus ensuring maximal sensitivity to these functional groups and minimal cross-reactivity with other chloroacetanilide herbicides. The chloroacetamide group would normally be expected to react readily under basic conditions with protein thiols. Preliminary experiments were performed with the aim of directly conjugating alachlor to proteins, and suggested that insufficient levels of conjugation (epitope density) would be obtained for efficient antibody production in immunized rabbits.

In order to increase the availability of protein thiols and therefore increase the maximum epitope density, we employed the protein thiolating agents S-acetylmercaptosuccinic anhydride (AMSA) (8) and N-acetyl-homocysteine thiolactone (AHT) (9). Such thiolating groups are thought to react with the epsilon amino groups of lysines in proteins. Thiolation of proteins, followed by thiolate displacement of the alachlor chlorine atom, generated the alachlor-IgG immunizing antigen and the alachlor-BSA coating antigen, with proposed structures as illustrated in Figure 1. The use of ^{14}C-labeled alachlor permitted facile verification of the covalent attachment of alachlor to proteins, as well as measurement of epitope density. NMR spec-troscopic techniques are currently under development in our laboratories for the purpose of structurally characterizing alachlor-protein conjugates, in order to supplement the information provided by radioanalytical techniques.

Since only the alachlor moiety was common to these two hapten-protein conjugates, antibodies generated using the alachlor-IgG conjugate and reacting with the alachlor-BSA conjugate could be considered as recognizing only the alachlor moiety and not any portion of the linking reagent or the protein carrier. The checkerboard assay demonstrated the presence of antibodies in the antisera which reacted with the coating antigen (alachlor-BSA). The inhibition ELISA was subsequently employed to demonstrate that alachlor was able to inhibit the binding of the antibodies to the alachlor-BSA conjugate, and to establish the affinity of the antibodies toward alachlor. The optimized alachlor inhibition ELISA was most effective in the range from 0.2 ppb to 8.0 ppb alachlor in water, with corresponding percentages of absorbance ranging from 80% to 10%. The percentages of absorbance for the seven alachlor standards were obtained from 20 separate assays conducted on different days and on different plates. The means and standard deviations were used to generate the standard curve shown in Figure 2. The percentage coefficient of variabilities (% CV) ranged from 4.2% at 0.2 ppb to 18.6% at 8.0 ppb.

Cross-Reactivity of Antibodies. Alachlor belongs to a family of structurally similar chloroacetanilide herbicides. Specificity of the antibodies for alachlor was therefore crucial for the successful application of this assay to environmental samples. The strategy we employed for covalently linking alachlor to proteins through the chloroacetamide group was expected to assure a high degree of specificity for the substitution pattern found in alachlor. To test this assumption, we conducted extensive studies to determine the specificity of the antibodies for alachlor. The alachlor reactivity was defined as 100.0%. As

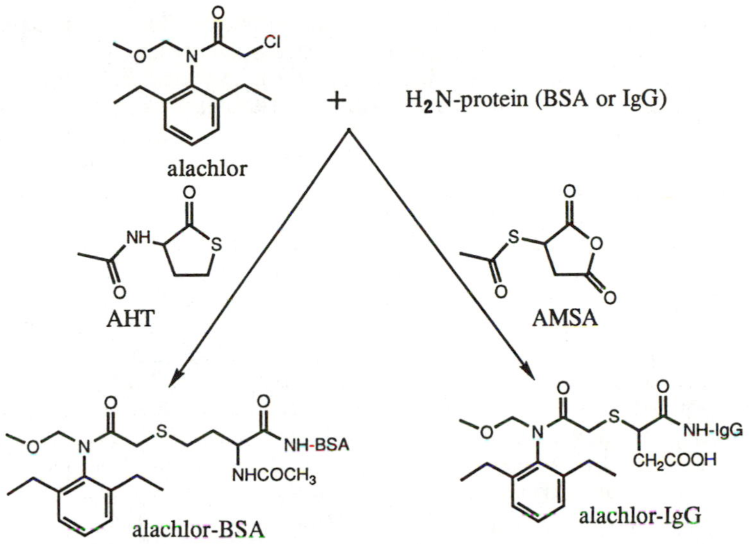

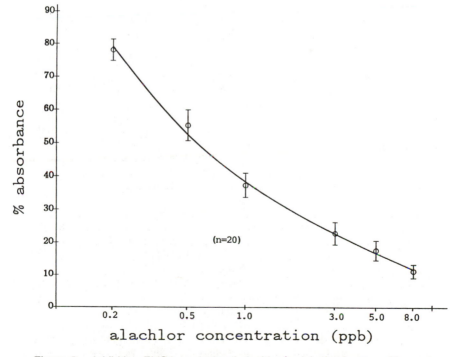

Figure 1. Covalent conjugation of alachlor to BSA with AHT, and to sheep IgG with AMSA. Alachlor-BSA was used as the coating antigen and alachlor-IgG was used as the immunizing antigen.

Figure 2. Inhibition ELISA analysis of alachlor in deionized water. Means and standard deviations were calculated from 20 separate analyses.

illustrated in Figure 3, the reactivity of the antibodies for acetochlor (2) and metolachlor (3) was 4.4% and 1.8% , respectively. Among other chloroacetanilide herbicides such as butachlor (4), amidochlor (5), and propachlor (6), none showed more than 2.0% cross-reactivity. Furthermore, absence of the N-methoxymethyl side chain of alachlor, as in the secondary chloroamide (7), resulted in a complete loss of antibody reactivity. Our results suggest that the reactivity of the antibodies is significantly affected by alterations in the methoxymethyl side chain.

The absence of the chlorine atom, as in nor-chloro alachlor (8), decreased the cross-reactivity 5-fold, suggesting that the chlorine atom also contributes significantly to the reactivities of the antibodies toward alachlor. Substitution of a hydroxyl group for chlorine (9), or the absence of both the methoxymethyl and chloroamide groups (10) resulted in almost complete loss of antibody cross-reactivity. These results confirmed that attachment to proteins via the chloroacetamide moiety through a thioether linkage is an effective strategy for designing immunizing antigens which elicit antibodies highly specific for the aniline ring and N-alkyl side chain of alachlor.

It was of particular interest to determine whether soil metabolites of alachlor would show significant antibody reactivity, since such metabolites might occur in environmental samples for which alachlor ELISA analysis could be used. The two major soil metabolites of alachlor have been identified as the oxanilic (11) and sulfonic acids (12) (10). The cross-reactivity of the antibodies for these compounds was found to be 2.5% and 2.3%, respectively (Figure 3).

We were also interested in observing cross-reactivities for alachlor analogues and metabolites bearing altered substitution patterns on the acetamide moiety. Since alachlor was conjugated to the immunizing antigen using intermediate sulfur-bearing groups, it was not surprising to find that some of the alachlor analogues showing the greatest cross-reactivity contained a thioether moiety (Figure 4). The greatest reactivity (188.0%), almost twice that of alachlor, was observed with methylsulfide (13). On the other hand, when the sulfur was oxidized as in sulfoxide (14) and sulfone (15), the cross-reactivity was reduced to 15.0% and 9.4%, respectively. The secondary amide sulfide (16) surprisingly produced no cross-reactivity. Some of the methylthio analogues (13-15) have previously been reported as products of animal metabolism of alachlor (10). Another commonly observed animal metabolite of alachlor is the mercapturic acid (17) (N-acetyl cysteine conjugate), which showed a 65.0% cross-reactivity with the antibodies in the present study. The high degree of cross-reactivity observed for 17 can be understood from the structural similarity of the mercapturic acid moiety of 17 with the thiolating reagents used for alachlor conjugation (Figure 1). The cross-reactivity for the mercapturate 17 was significantly reduced with the oxidation of the sulfur atom (18) or in the absence of the methoxymethyl group (19). A different thioether conjugate of alachlor was the glutathione conjugate (20), which showed a cross-reactivity of 27.5%.

Our cross-reactivity studies demonstrated that the antibodies were sensitive to modifications in the methoxymethyl side chain of alachlor, and were able to distinguish alachlor from structurally similar chloroacetanilide herbicides. Minor modifications in the N-methoxymethyl side chain of alachlor

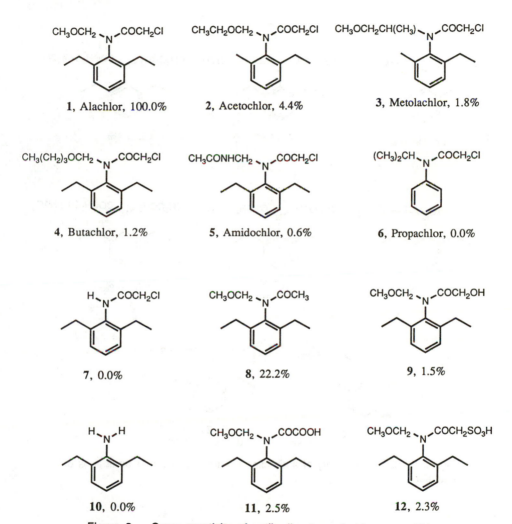

Figure 3. Cross-reactivity of antibodies toward chloroacetanilide herbicides and analogues of alachlor.

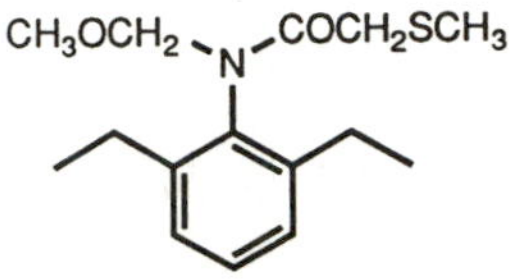

13, 188.0%

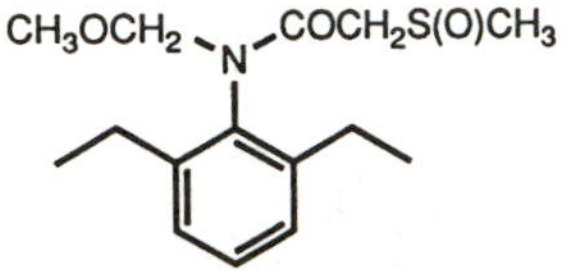

14, 15.0%

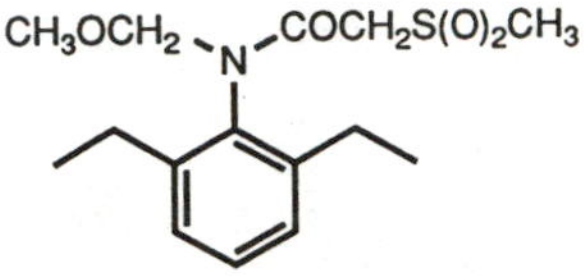

15, 9.4%

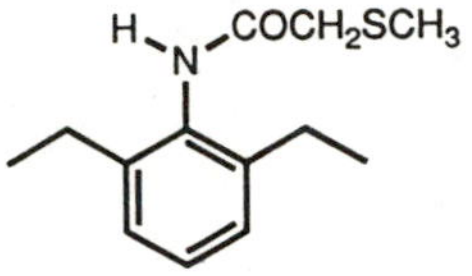

16, 0.0%

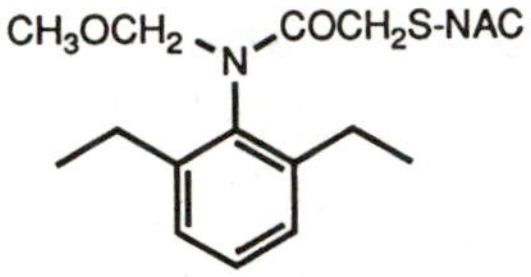

17, 65.0%
(NAC= N-acetyl cys)

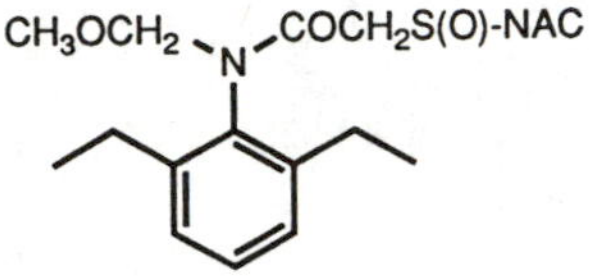

18, 12.8%

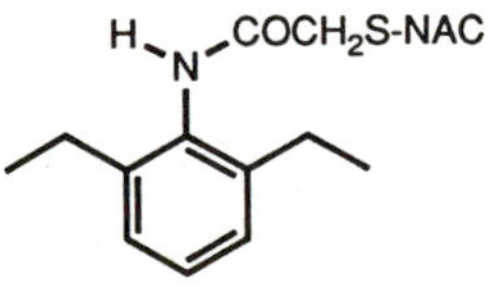

19, 0.0%

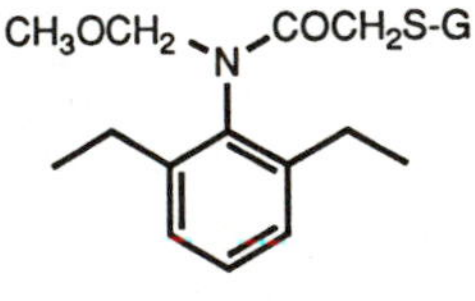

20, 27.5%
(G= glutathione)

Figure 4. Cross-reactivity of antibodies toward the thioether analogues of alachlor.

led to significant reductions in reactivity. The presence of either sulfur or chlorine at the acetamide 2-position of the alachlor molecule was important for the reactivity, while oxidation of the sulfur in the thioether linkage reduced the cross-reactivity substantially. Minimal cross-reactivity was observed for soil metabolites of alachlor. These results indicated that the antibodies could be used for environmental alachlor analysis without significant interference from other related compounds.

ELISA Analysis of Environmental Water Samples. The concentration of alachlor in environmental surface water is generally negligible, but can vary depending on patterns of herbicide use and on geographical and environmental factors. Under worst-case conditions following planting, alachlor concentrations of less than 20 ppb have been reported for river water samples from Ohio (11). On the other hand, most of the water samples analyzed in the same study during the period from May to August showed non-detectable levels of alachlor.

A potential application of our ELISA method would be for the analysis of alachlor in water. To test this possibility, 208 water samples were collected for analysis from rivers and water treatment plants. Some of the samples were intentionally spiked with alachlor as controls. The samples were analyzed by ELISA without any pretreatment, and by an established GC/MS method. For the ELISA analysis, a sample size of 1.0 mL was required. The GC/MS analysis, on the other hand, required 1.0 L of sample volume. The results of the ELISA and GC/MS analyses are presented in Figure 5. The X axis displays ppb of alachlor as determined by ELISA, and the Y axis displays ppb of alachlor as determined by GC/MS. The correlation coefficient of the two methods was 0.84, and the slope of the regression line was 0.74.

Analysis of spiked samples showed that ELISA was generally less accurate and less precise than the GC/MS method. The percentage coefficient of variabilities for ELISA ranged from approximately 10% to 40%, and were considerably higher than the GC/MS analysis. The samples ranged from river water containing suspended soil particulates to finished water from water treatment plants. Since ELISA is conducted without any sample pretreatment, the assay is more susceptible to sample matrix effects. We suggest that the higher variability of the ELISA data contributes to the observed discrepancy between the ELISA and GC/MS methods.

However, the ELISA method can be used very effectively as a primary screen to select water samples with reference to an arbitrary threshold of alachlor concentration. A sample selected by ELISA and confirmed by GC/MS as being below a threshold can be classified as a correct negative. Correspondingly, a sample selected by ELISA and confirmed by GC/MS as being equal to or above a threshold can be classified as a correct positive. Threshold levels were chosen at 0.5, 1.0, and 5.0 ppb.

Using the 1.0 ppb threshold as an example (Table I), 145 out of 147 samples selected by ELISA to contain below 1.0 ppb of alachlor were confirmed by GC/MS, which translated to a 99% incidence of correct negatives. On the other hand, only 29 out of 61 samples selected by ELISA to contain greater than or equal to 1.0 ppb of alachlor were confirmed by GC/MS, thus resulting in a correct positive incidence of only 48%. Dividing the number of ELISA positive samples (61) by the total number of samples (208) produced the percentage

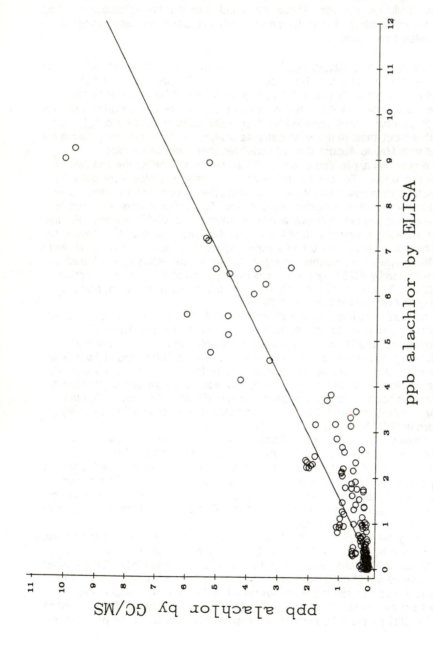

Figure 5.   Analysis of alachlor in environmental water samples by ELISA and by GC/MS. The correlation coefficient was 0.84 based on the analysis of 208 water samples.

(29%) of samples requiring confirmation by GC/MS. The use of ELISA as a screen would thus have reduced the sample load for GC/MS by 71%. The use of a 5.0 ppb threshold in ELISA would have reduced the sample load for GC/MS by 94%. The very high percentage of correct negatives demonstrate that ELISA can be used reliably as a screen. However, because of the comparatively low percentage of correct positives, positive samples selected by ELISA will require confirmation by an alternative method.

Table I. ELISA as a primary screen for the analysis of alachlor in environmental water samples

Threshold (ppb)	% Correct negatives	% Correct positives	% Samples to GC/MS
0.5	117/122=96	54/86=63	86/208=41
1.0	145/147=99	29/61=48	61/208=29
5.0	193/195=99	7/13=54	13/208= 6

In conclusion, our results demonstrate an effective application of polyclonal antibodies directed toward alachlor. These antibodies were used to develop an ELISA with a detection range from 0.2 to 8.0 ppb for alachlor in water. The demonstrated ability of the antibodies to distinguish alachlor from other chloroacetanilide herbicides allowed the successful application of this assay to the analysis of environmental water samples. Based on the analysis of a limited number of environmental water samples, the ELISA method was shown to be less accurate and precise than GC/MS. On the other hand, ELISA offers considerable advantages over GC/MS in terms of cost, speed, sample through-put, and portability. Our results suggest that the most effective use of this ELISA would be as a primary screen, in which samples falling below a threshold level of alachlor can be rapidly and cost-effectively eliminated from further instrumental analysis.

Acknowledgment

We are grateful to J. M. Malik for valuable suggestions and comments. We thank R.G. Smith and F.M. Triebe for GC/MS analysis of water samples.

Literature Cited

1. Hammock, B.D.; Mumma, R.O. In Pesticide Analytical Methodology; Harvey, J., Zweig, G. Eds.; American Chemical Society Symposium Series No 136, Washington D.C., 1980, pp. 321-352.
2. Van Emon, J.M.; Seiber, J.N.; Hammock, B.D. In Bioregulators for Pest Control; American Chemical Society Symposium Series No 276, Washington D.C., 1985, pp. 307-316.

3. Hammock, B.D.; Gee, S.J.; Cheung, P.Y.K.; Miyamoto, T.; Goodrow, M.H.; Van Emon, J.; Seiber, J.N. In Pesticide Science and Biotechnology; Greenhalgh, R., Roberts, T.R., Eds.; Blackwell Scientific: Ottawa, 1987; pp 309-316.
4. Jung, F.; Gee, S.J.; Harrison, R.O.; Goodrow, M.H.; Karu, A.E; Braun, A.L.; Li, Q.X.; Hammock, B.D. Pestic. Sci. 1989, 26, 303-317.
5. Wratten, S.J., Feng, P.C.C. In Development and Applications of Immuno-assays or Food Analysis; Rittenburg, J.H. Ed.; Elsevier:New York,1990, pp. 201-220.
6. Feng, P.C.C.; Wratten, S.J.; Horton, S.R.; Sharp, C.R.; Logusch, E.W. J. Agric. Food Chem. 1990, 38, 159-163.
7. Campbell, A.M. In Monoclonal Antibody Technology: The Production and Characterization of Rodent and Human Hybridomas: Burdon, R.H., Knip-penberg, P.H., Eds.; Elsevier: New York, 1984.
8. Klotz, I.M.; Heiney, R.E. Archiv. Biochem. Biophys. 1962, 96, 605-612.
9. Singer, S.J.; Fothergill, J.E.; Shainoff, J. J. Amer. Chem. Soc. 1960, 82, 565-571.
10. Sharp, D. B. In Herbicides: Chemistry. Degradation and Mode of Action; Kearney, P. C., Kaufman, D. D., Eds.; Dekker: New York, 1988, pp. 301-333.
11. Baker, D.B. J. Soil Water Conservation 1985, 40, 125-132.

RECEIVED May 8, 1990

Chapter 16

Competitive- and Inhibition-Type Immunoassay for Determination of Endosulfan

Bernhard Reck and Jürgen Frevert

Batelle-Institut e.V., Am Römerhof 35, 6000 Frankfurt am Main 90, Federal Republic of Germany

Two different types of immunoassays, a competitive-type and an inhibition-type, were developed and compared for the insecticide, endosulfan. The detection range of both assay types was similar, 3-500 ng/ml for the competitive-type and 5-500 ng/ml for the inhibition-type assay. Metabolites of endosulfan and other insecticidal chloro-hydrocarbons possessing a hexachlorocyclopentene structure exhibited considerable crossreaction in both assays, so these immunoassays could be used for detection of class-specific compounds. The inhibition-type immunoassay showed less susceptibility to interfering factors and, therefore, seemed to be more suitable for environmental analysis.

Conventional analyses for the insecticide, endosulfan, in environmental samples requires time-consuming multistep cleanup procedures prior to the final quantification by gas chromatography ([1]). For this reason, a sensitive and rapid ELISA could be of great value if large numbers of samples have to be analyzed. The advantages of immunoassays for residue analysis have been described by Hammock and Mumma ([2]).

To develop such an immunological assay for endosulfan, specific antibodies directed against endosulfan are necessary. As a low molecular weight substance, endosulfan per se is not immunogenic, i.e. it is not capable of inducing the synthesis of antibodies in animals. The first step in the ELISA development is, therefore, the covalent coupling of endosulfan or a derivative to a carrier protein. Therefore, a derivative of endosulfan, endosulfandiol, was linked to a high molecular weight protein, Keyhole limpet hemocyanin (KLH) by first synthesizing the hemisuccinate and then the N-hydroxysuccinimid

0097–6156/90/0442–0193$06.00/0

ester. This active ester readily reacts with free amino groups of the KLH (3), producing an immunogen which was then used to raise specific antibodies in rabbits.

For the competitive-type ELISA (description, see below), endo-sulfan or a derivative must also be conjugated to an enzyme. To avoid the antibodies from reacting with the linkage between endosulfandiol and KLH, another coupling procedure was chosen. Endosulfandiol was linked to ethylenediamine utilizing carbodiimidazole such that a free amino group was available. The vicinal hydroxy groups of the glycoprotein enzyme, horseraddish peroxidase, were oxidized with $NaIO_4$, resulting in the formation of aldehyde groups (4). These aldehyde groups then reacted with the free amino group of the endosulfandiol coupled to the ethylenediamine to produce the enzyme conjugate. Because of the strongly hydrophobic character of endosulfandiol, the synthesis of the enzyme conjugate was the most critical step in the development of the immunoassay. When horseraddish peroxidase was linked to several endosulfan derivatives per molecule, it became very insoluble and lost its enzymatic activity. An enzyme conjugate with a ratio of 1:1 was most favorable for the immunoassay. On the other hand, a KLH-endosulfan conjugate with a high number of endosulfan per KLH was insoluble and was an effective immunogen producing a high antibody titer. Using these reagents, the inhibition and the competitive type immunoassay were established and their characteristics are compared in Table I.

Table I. Comparison of Competitive- and Inhibition-Type
Immunoassay

	Competitive-type immunoassay	Inhibition-type immunoassay
incubation steps	1	2
time required	1.5 hr	>5 hr
detection limit	3 ng/ml	5 ng/ml
detection range	3-500 ng/ml	5-500 ng/ml
consumption of antisera	high	low
stability of reagents	peroxidase-endosulfan very unstable	no unstable reagent
susceptibility to interfering factors in environmental samples	high (inhibition of enzyme activity)	low (inhibition of antibody-antigen binding)

In the competitive-type immunoassay, the antibody was immobilized in the wells of a polystyrene microtiter plate (5). The sample to be analyzed, which potentially contained endosulfan, was then incubated with a fixed amount of peroxidase-endosulfandiol conjugate. Free endosulfan and the endosulfandiol enzyme conjugate competed for binding to the immobilized antibody. The wells were then washed and the chromogen, o-phenyldiamine, was added. The optical density of the color of the solution in the wells was determined. The intensity of the color was proportional to the amount of bound peroxidase conjugate. A standard curve derived from the analysis of a serial dilution of endosulfan was used for the basis of analysis of samples. Figure 1 gives an example of a standard curve and demonstrates how variation of the enzyme-conjugate influences the measuring range of the assay. For comparison of different test runs, the A/A_{max} ratio was plotted versus endosulfan concentration (A_{max} was the optical density of the test solution without the analyte). The detection limit of the assay was lowered by reducing the concentration of the peroxidase conjugate. On the other hand, lowering the concentration of the enzyme conjugate reduced the absorbance which resulted in a decreased sensitivity. Therefore, one had to identify the optimal concentration for the enzyme conjugate. A peroxidase conjugate concentration of 150 ng/ml was used to achieve a sufficiently high absorbance. The final competitive-type ELISA resulted in measuring endosulfan concentrations in the range of 3-500 ng/ml.

A major drawback of competitive-type immunoassay arose when environmental samples were analyzed. In this procedure, the enzyme conjugate was added directly to the sample, thus, all substances in the sample which have an effect on the enzyme would interfere with the assay. This dramatically limits the use of the competitive-type ELISA assay. Therefore, an inhibition-type ELISA was developed which would supposedly be less susceptible to interfering factors present in the sample. In this type of assay, a protein-endosulfandiol conjugate is immobilized in a microtiter plate. In the first step, a limited, constant amount of antibody was incubated with the endosulfan containing sample. This can be done in the microtiter plate or during an external incubation step, i.e., sample and antibody were preincubated in a tube. Antibodies which had not reacted with endosulfan would bind to the immobilized endosulfandiol conjugate on the test plate. The amount of bound antibody, which was dependent on the amount of endosulfan present in the sample, was then quantitated with a second, enzyme-labeled antibody directed against the first antibody. Because of the washing steps between the incubations, the enzyme is not effected by potentially interfering substances present in the sample.

Standard curves for different incubation procedures are shown in Figure 2. Only minor differences in sensitivity were found and this allows a variable test setup in respect to preincubation. No appreciable differences were demonstrated between the competitive- and inhibition-type immunoassays (Fig 2). In contrast, differences in the crossreaction with endosulfan derivatives relative to the two type assays were found. Because the coupling of an endosulfan derivative was achieved by using a hydroxyl group of endosulfandiol, the hexachloropentene structure was exposed, and a strong crossreactivity with all compounds having this structure could be anticipated.

Table II shows that all derivatives of endosulfan (except endosulfanlactone in the competitive-type immunoassay) exhibited a significant affinity to the antiserum. However, there were major differences in the amount of crossreaction in the assay types, e.g., endosulfandiol was better recognized in the competitive-type immunoassay than in the inhibition-type assay. A possible reason for these differences may be that in one case the antibodies were immobilized and the binding of these antibodies to the solid phase could have an effect on the avidity.

Table II. Crossreaction of Endosulfan Derivatives

Pesticide or Derivative	% Crossreaction	
	Inhibition-type	Competitive-type
α–Endosulfan	100	100
β-Endosulfan	128	100
Endosulfandiol	34	158
Endosulfanether	140	57
Endosulfansulfate	131	203
Endosulfanlactone	11	0.02

The behavior of other insecticidal chlorohydrocarbons has also been investigated, and the crossreactions of these are presented in Table III. All compounds possessing a hexachlorocyclopentene structure showed a significant reactivity in both assay types. Surprisingly, both immunoassay had a higher sensitivity for endrin than for endosulfan. The stereochemical analogue of endrin, dieldrin, however, showed only moderate affinity to the antibody. Because of this crossreaction, neither type of assays, developed with this antiserum, allows for specific detection of endosulfan itself, but can be used for screening for the presence of a class of chlorohydrocarbons.

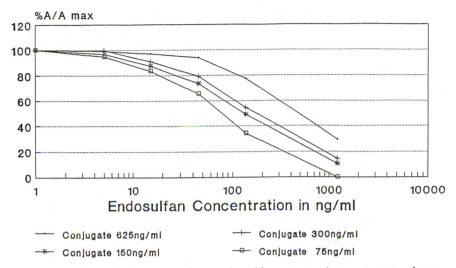

Figure 1. Standard curves for endosulfan at varying concentrations of endosulfandiol-horseradish peroxidase conjugate.

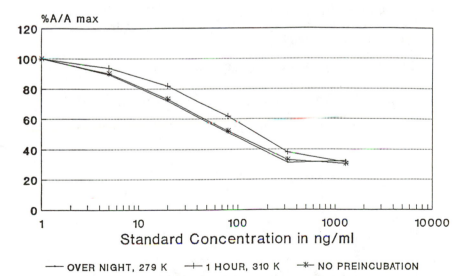

Figure 2. Standard curves for endosulfan at varying times of the antibody-sample preincubation step.

Table III. Crossreaction of Insecticidal Chlorohydrocarbons Correlated
to α-Endosulfan (100%)

Pesticide	% Crossreaction	
	Inhibition-type	Competitive-type[a]
Aldrin	14	16
Alodan	6	7
Heptachlor	12	n.d.
Heptachloroepoxide	20	n.d.
Endrin	364	180
Dieldrin	41	n.d.
Lindane	2	3
Kelevan	<0.01	n.d.

[a] n.d. = not determined.

Literature Cited

1. Zweig, G.; Archer, T.E. J. Agric. Food Chem. 1960, 8, 190-192.

2. Hammock, R.R.; Mumma, R.O.   In Pesticide Analytical
    Methodology; Harvey, I., Zweigh, G., Eds., American Chemical
    Society: Washington, DC, 1980; p 321.

3. Tijssen, P.; Kurstak, E. Anal. Biochem. 1984, 136, 451-457.

4. Dreher, R.M.; Podratzki, B. J. Agric. Food Chem. 1988, 36, 1072.

5. Engvall, E.; Perlman, P. Immunochemistry 1971, 8, 871-874.

RECEIVED July 26, 1990

Chapter 17

Monoclonal Antibody-Based Enzyme Immunoassay for Atrazine and Hydroxyatrazine

Jean-Marc Schlaeppi, Werner Föry, and Klaus Ramsteiner

Pharmaceuticals Research Laboratories and Agricultural Division, Ciba–Geigy, Ltd., R–1056.3.19, 4002 Basel, Switzerland

Monoclonal antibodies were obtained against atrazine and its metabolite hydroxyatrazine by immunizing mice with atrazine or hydroxyatrazine protein conjugates. By competitive ELISA, we observed that the antibodies raised against hydroxyatrazine cross-reacted mainly with hydroxypropazine. The antibodies raised against atrazine cross-reacted with propazine, prometone, prometryne, and to a much lower extent with a few other s-triazines and hydroxy-s-triazines. Atrazine could be detected in water samples down to 50 ppt. Average recoveries measured by ELISA from soil samples fortified with atrazine or hydroxyatrazine were comparable to those measured by GLC or HPLC. Soil samples of unknown atrazine content were analyzed by GLC, GC-MS, and by ELISA. The results show that the ELISA immunoassay represents a valuable detection method for trace amounts of atrazine and hydroxyatrazine in soil.

Triazines such as atrazine, propazine or simazine, are widely used herbicides. Currently, the determination of atrazine in water and soil samples is mainly done by GLC (1,2) or HPLC (3-5). However, these procedures require cumbersome cleanup steps which could be avoided by using immunoassays as an alternative approach to residue analysis. Such immunochemical determination based on competitive binding of herbicides or pesticides to an antibody (6,7), has been described recently for s-triazines (8-10). To develop our immunoassay for atrazine and the important nonphytotoxic metabolite hydroxyatrazine, we took advantage of hybridoma technology to obtain monoclonal antibodies (MAbs), which allowed a better definition of the specificity of the assay, as well as an unlimited supply of reagents.

Experimental Section

Synthesis of Atrazine and Hydroxyatrazine Conjugates. Small molecules like atrazine become immunogenic in animals only upon conjugation to a carrier protein. Therefore, two derivatives of atrazine were synthesized, which contained a spacer ending with a carboxyl group for attachment to the protein (Figure 1). These derivatives were converted to the hydroxy analogues by adding 6N HCl at room temperature. The N-hydroxysuccinimide esters of these derivatives were prepared by reacting them with N-hydroxysuccinimide and N,N'-dicyclohexylcarbodiimide (11), and they were covalently coupled to either bovine serum albumin (BSA) or keyhole limpet hemocyanin (KLH). The molar ratio of atrazine and hydroxyatrazine to BSA was determined by UV absorbance spectrophotometry. It varied between 9/1 and 11/1.

Immunization and Fusion Protocol. Groups of five BALB/c female mice (4-6 weeks old), were given series of three injections every two weeks, with KLH-conjugated atrazine or hydroxyatrazine (50 ug/injection) mixed with Freund's adjuvant. After a rest period of two months, the mice were boosted with 500 ug of the conjugate. Three to four days later, the mice were sacrificed and the spleen cells were fused with the murine myeloma cell line Sp 2/0.Ag14 (12,13). The positive hybridomas were cloned and expanded in mice and the MAbs were purified from the ascitic fluid by ammonium sulfate precipitation, and DEAE-cellulose anion-exchange chromatography (14).

Competitive ELISA for Atrazine or Hydroxyatrazine Determination. The two-step competitive ELISA using the BSA-conjugate to coat the 96-well microtiter plates was done as described elsewhere (15). Inhibition curves were obtained by plotting B/Bo x 100 (percent bound) vs. the concentration of inhibitor present (Bo represented the absorbance measured without s-triazines inhibitor added to the antibody, and B, the absorbance measured with various concentrations of s-triazines inhibitor). I_{50} represented the concentration of atrazine or analogues, reducing the ELISA signal to 50% of the control. I_{50} was calculated using an adaptation of the curve fitting program ENZFITTER (R.J.Leatherbarrow, Elsevier-Biosoft) based on a four-parameter logistic curve (16).

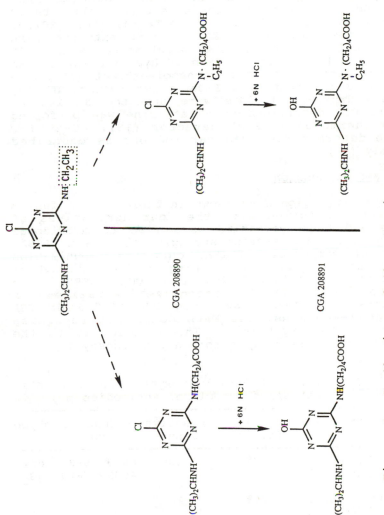

Figure 1. Synthesis of the atrazine and hydroxy-atrazine derivatives used for coupling the haptens to the carrier proteins.

Quantitative measurements of atrazine or hydroxyatrazine in soil extracts or water samples, by ELISA, were calculated using the same program, the curve fitting being based on standards run on each microtiter plate.

Soil Samples Analysis. Standard soil samples from various locations were used for this study. Aliquots (2 g) were extracted in 20 ml of methanol/water (80/20: v/v). For the competitive ELISA, soil extracts were routinely diluted 1:40 in PBS supplemented with 0.1% Tween-20. The HPLC determination of hydroxyatrazine was done after cleanup of the methanol-extract (17). The samples were injected in a Lichrospher column, SI 60, and the hydroxy-s-triazines were detected at 240 nm (17). The GLC determination of atrazine was performed using a thermoionic (P-N) detector (18). GC-MS for atrazine determination was carried out as described previously (19).

Results and Discussion

MAbs Characterization. As shown in Table I, all fusion experiments performed with the four derivatives of atrazine and hydroxyatrazine gave rise to hybridomas which secreted MAbs reacting strongly by direct ELISA. However, by competitive ELISA, only the immunization with the two derivatives containing the [(1-carboxybut-4-yl)amino] side chain (GCA 208891 and CGA 208890) gave rise to MAbs which recognize atrazine or hydroxyatrazine, whereas by using CGA 208895 or CGA 208894 as immunogens, the MAbs could not recognize atrazine nor hydroxyatrazine. They bound only to the derivatives and were useless for an immunoassay.

Table I.   Production of Anti-hydroxyatrazine and Anti-atrazine Monoclonal Antibodies

Fusion No.	Immunogen	Fusion Efficiency	Positive Clones	MAbs No.	Isotype
4009	CGA 208891 (OH-Atr.)	97%	10	4009-85-3 4009-77-20	IgG_1 IgG_1
4010	CGA 208895 (OH-Atr.)	82%	10	--	
4063	CGA 208890 (Atr.)	74%	5	4063-21-1	IgG_1
4006	CGA 208894 (Atr.)	31%	1	--	

The 10 anti-hydroxyatrazine MAbs could be distributed into two groups, based on their cross-reactivity pattern obtained by competitive ELISA. In the first group (represented only by the MAb 4009-85-3), the cross-reactivity was restricted mainly to hydroxypropazine (Table II), whereas in the second group (represented by the MAb 4009-77-20), the cross-reactivity encompassed other hydroxy-s-triazines (Table II). In both groups, no cross-reactivity with active s-triazines (such as atrazine) was observed, ie. the binding of the MAbs was restricted to the presence of a hydroxyl group in the triazine ring. The minimum detectable amount of hydroxyatrazine, defined as being the concentration required to produce a decrease in the percent of MAb bound equal to two times the standard deviation of the blank response, was 0.1 and 0.05 ng/ml for MAb 4009-85-3 and 4009-77-20 respectively (average of 15 determinations). In the case of mice immunized with the atrazine-conjugate, all 5 MAbs obtained from one fusion experiment showed a comparable pattern of cross-reactivity as represented by clone 4063-21-1 (Table II). In this case, the MAbs showed high cross-reactivity with propazine, prometryne and prometone and cross-reacted to a much lower extent with a few other s-triazines and hydroxylated metabolites (Table II). For MAb 4063-21-1, the limit of detection for atrazine in buffer was 0.05 ng/ml.

It follows that an immunoassay for quantitative determination of atrazine, based on the use of MAb 4063-21-1, would not be fully specific; the ELISA values representing a compound response to all cross-reacting substances. One way to improve the specificity would be to combine the use of MAb 4063-21-1 with that of the anti-hydroxyatrazine MAb 4009-85-3, allowing a clear distinction between the response due to s-triazines and that of cross-reacting hydroxylated metabolites. However, the immunoassay would still not discriminate between atrazine and a few other triazines.

Soil Extract and Soil Fortification Experiments. We tested for possible interfering soil coextracted materials by adding hydroxyatrazine to the methanol/water extract of five soils of known composition. Hydroxyatrazine was measured by ELISA using MAb 4009-85-3 before and after the addition of hydroxyatrazine to the soil extract. As shown in Figure 2, all soil types tested gave acceptable recoveries (97% mean recovery), suggesting that in these samples, no interfering substances were coextracted.

In the second type of experiment, five soil samples without traces of hydroxyatrazine nor hydroxypropazine as measured by HPLC, were fortified with both metabolites in equal concentration. After the methanol extraction, they were analyzed by ELISA and HPLC. The

TABLE II. Cross-Reactivity of Various S-Triazines Analogues with One Anti-Atrazine MAb (MAb 4063-21-1) and Two Anti-Hydroxyatrazine MAbs (MAb 4009-85-3 and MAb 4009-77-20)

Inhibitor	MAb 4063-21-1		MAb 4009-85-3		MAb 4009-77-20	
	I50 (ng/ml) (a)	% Cross-Reactivity (b)	I50 (ng/ml)	% Cross-Reactivity (c)	I50 (ng/ml)	% Cross-Reactivity (c)
Atrazine	0.45	100.0	>1000	<0.1	>1000	<0.05
Propazine	0.5	90.0	>1000	<0.1	>1000	<0.05
Simazine	18	2.5				
Desmetryne	90	0.5				
Terbuthylazine	9	5.0				
Trietazine	4	11.3				
Ametryne	9	5.0	>1000	<0.1	>1000	<0.05
Atratone	7	6.4	>1000	<0.1	>1000	<0.05
Prometryne	0.8	56.3				
Simetryne	27	1.7				
Prometone	0.6	75.0				
Metribuzin	>1000	<0.05				
Metamitron	>1000	<0.05				
OH-Atrazine	21	2.1	0.95	100.0	0.5	100.0
OH-Propazine	5.5	8.2	0.95	100.0	0.6	83.3
OH-Simazine	180	0.3	65	1.5	1.2	41.7
OH-Desmetryne	360	0.1	8	11.9	0.7	71.4
OH-Terbuthylazine	8	5.6	19	5.0	3.3	15.2
G30033	52	0.9	>1000	<0.1	250	0.2
GS17794	250	0.2	29	3.3	4	12.5
G28521	>1000	<0.05	>1000	<0.1	>1000	<0.05

a) Inhibitor concn reducing the ELISA signal to 50% of the control.
b) (atrazine concn for 50% inhibn / s-triazine analogue concn for 50% inhibn) x 100.
c) (hydroxyatrazine concn for 50% inhibn / s-triazine analogue concn for 50% inhibn) x 100.

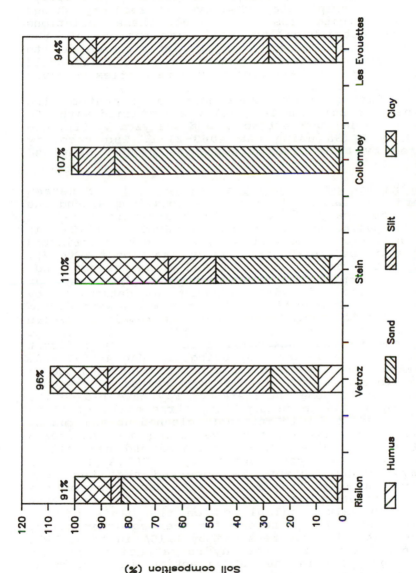

Figure 2. Percentage of added hydroxyatrazine (100 ppb) recovered from fortified soil extracts. The ELISA determination was performed with MAb 4009-85-3.

recoveries determined by ELISA (60% mean recovery) were comparable to those determined by HPLC (64% mean recovery) (Figure 3). The intra- and interassay coefficients of variation obtained by ELISA were 6.3% (23 assays) and 14% (29 assays) respectively (all assays were done in triplicate). The overall recovery showed variations ranging from 46 to 100%. These variations could be attributed to soil adsorption of the hydroxytriazines. Indeed, possible interference due to the high organic material content in these soils could explain the greater variation of the recoveries observed both by HPLC and ELISA.

The fortification experiment was repeated with atrazine. One soil sample (#15) was fortified with 100 ppb atrazine and then extracted and analyzed by GLC, GC-MS and by ELISA using MAb 4063-21-1. The recovery measured by all three methods was almost complete (>90%).

Determination of Hydroxyatrazine in Soil. Nineteen standard soil samples from various locations around the world were analyzed for their hydroxyatrazine and hydroxypropazine content either by HPLC or ELISA. As shown in Figure 4, the soil samples could be distributed into two groups. In the first group (samples # 1 to 7), a good correlation between HPLC and ELISA was obtained. The soil sample # 1 contained a high amount of hydroxyatrazine and hydroxypropazine as determined by both methods, whereas the six other samples were devoid of these metabolites. However, for the remaining twelve HPLC positive soil samples (# 8 to 19), the ELISA values were close to the background level. No significant differences were observed by using the MAb 4009-77-20. Confirmatory test by GC-MS could not be performed due to the low volatility of the hydroxytriazines. Therefore, to test whether some interferences may have biased the monitoring at 240 nm during HPLC, three soils (# 17 to 19) were extracted with methanol, cleaned up and run by HPLC (17). The putative hydroxyatrazine peak monitored at 240 nm was collected, concentrated and neutralized before being analyzed again by ELISA for its hydroxyatrazine content. We observed that the HPLC fraction obtained from sample # 17 contained only a small amount of hydroxyatrazine, representing less than 10% of the values calculated on the basis of the absorbance at 240 nm, whereas for the two other soils, no hydroxyatrazine was measured by ELISA in these HPLC fractions. Moreover, the hydroxyatrazine content measured by ELISA in the HPLC fraction was consistent with the one measured in an aliquot of the methanol extract taken before cleanup. Very likely, the discrepancies between HPLC and ELISA were due to some interferences which biased the monitoring at 240 nm.

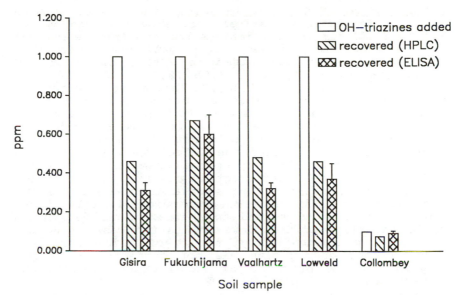

Figure 3. Recovery of hydroxytriazines from fortified soils. Comparison between HPLC and ELISA performed with MAb 4009-85-3.

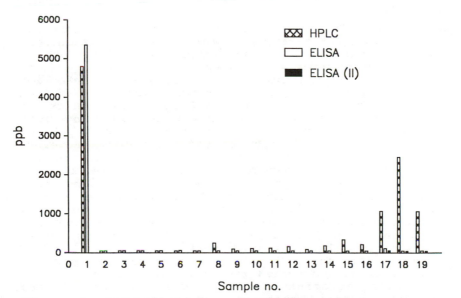

Figure 4. Determination of hydroxyatrazine and hydroxypropazine in soil samples. Comparison between HPLC and ELISA (performed with MAb 4009-85-3). For three samples, the HPLC fraction containing the hydroxyatrazine was analyzed again by ELISA (ELISA II) (see the text).

Determination of Atrazine in Soil. The   soil samples
analyzed for hydroxytriazines were also analyzed by GLC
and by ELISA for their triazines content (Figure 5). Out
of the nineteen soil samples, eight were essentially
devoid of s-triazines based on either ELISA or GLC
determination (# 4,7,8,9,13,15,18,19), whereas five
other samples (# 5,10,11,12,17) contained s-triazines
based on the GLC determination, but scored negative by
ELISA. Four of these soils were further analyzed by GC-
MS for confirmatory results. It appeared that no
atrazine was detected by this method, suggesting that
some interferences biased the detection of atrazine by
GLC. On the other hand, three soil samples (# 2,14,16)
were negative by GLC but slightly positive by ELISA. The
presence of cross-reacting s-triazines analogues may
very likely explain these discrepancies. Finally, three
samples (#1,3,6) contained some atrazine as determined
by either ELISA or GLC. However, in the case of sample
#1, which contained a high amount of hydroxyatrazine,
the presence of atrazine was not confirmed by GC-MS,
suggesting the presence of interfering coextracted
materials not yet identified.

Conclusion

We obtained stabilized hybridoma cell lines secreting
MAbs specific for atrazine, a widely used herbicide, and
for hydroxyatrazine, an important metabolite. As we
observed previously with water samples (15), a good
correlation was obtained between the current detection
method (HPLC or GLC) and a MAb based immunoassay (ELISA)
when fortified soils were analyzed. The limits of
detection of atrazine and hydroxyatrazine by both
methods were comparable. They corresponded to 50 ppb for
soil samples and 0.05 ppb for water samples. However,
by evaporating the methanol soil extracts before the
ELISA, the limit of detection in soil samples could be
reduced down to the ppb level. The analysis of undefined
soil samples with respect to their herbicide content
showed that for atrazine and hydroxyatrazine, some
discrepancies were observed between the two methods due
to biased detection during HPLC and GLC measurements.
For atrazine this was confirmed by additional GC-MS
analysis. Therefore, the ELISA immunoassay represents a
valuable detection method for trace amounts of atrazine
and hydroxyatrazine in soil, despite its limited
specificity due to cross-reacting substances.

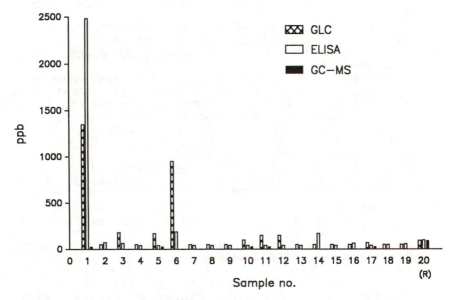

Figure 5. Determination of atrazine, propazine and simazine in soil extracts. Comparison between GLC, GC-MS and ELISA (performed with MAb 4063-21-1). (R = recovery).

Acknowledgments

We thank C. Erni and M. Lisibach for skillful technical assistance and Drs. G. Rosenfelder and W.D. Hörmann for their support throughout this work.

Literature Cited

1. Sirons, G. J.; Frank, R.; Sawyer, T. J. Agric. Food Chem. 1973, 21, 1016-1020.
2. Lee, H-B.; Chau, A.S.Y. J. Assoc. Off. Anal. Chem. 1983, 66, 651-658.
3. Vickrey, T.M.; Karlesky, D.L.: Blackmer, G.L. J. Assoc. Off. Anal. Chem. 1980, 63, 506-510.
4. Owens, D.S.; Sturrock, P.E. Anal. Chim. Acta 1986, 188, 269-274.
5. Ramsteiner, K.A. J. Chromatogr. 1989, 465, 410-416.
6. Hammock, B.D.; Mumma, R.O. In Pesticide Identification at the Residue Level; Gould, R.F., Ed.; American Chemical Society: Washington, DC., 1980; ACS Symp. Ser. No. 136, p 321-352.
7. Vanderlaan, M.; Watkins, B.E.; Stanker, L. Environ. Sci. Technol. 1988, 22, 247-254.
8. Huber, S. Chemosphere, 1985, 14, 1795-1803.
9. Dunbar, B. D.; Niswender, G. D.; Hudson, J. M. US Patent # 4 530 786, 1985.
10. Bushway, R. J.; Perkins, B.; Savage, S. A.; Lekousi, S. J.; Ferguson, B. S. Bull. Environ. Contam. Toxicol. 1988, 40, 647-654.
11. Kulkarni, N. P.; Blair, A. H.; Ghose, T. I. Cancer Res. 1981, 41, 2700-2706.
12. Galfre, G.; Howe, S. C.; Milstein, C.; Butcher, G. W.; Howard, J. C. Nature 1977, 266, 550-552.
13. Shulman, M.; Wilde, C.D.; Köhler G. Nature 1979, 276, 269-270.
14. Campbell, A.M. "Monoclonal Antibody Technology". In Laboratory techniques in Biochemistry and Molecular Biology, Burdon, R. H.; Knippenberg, P.H. Eds.; Elsevier, Amsterdam, 1984; Vol. 13, p 120-184.
15. Schlaeppi, J.-M.; Föry, W.; Ramsteiner, K. J. Agric. Food. Chem. 1989, 37, 1532-1538.
16. Raab, G. M. Clin. Chem. 1983, 29, 1757-1761.
17. Ramsteiner, K. A.; Hörmann, W. D. J. Agric. Food Chem. 1979, 27, 934-938.
18. Ramsteiner, K. In Methodensammlung zur Rückstandsanalytik von Pflanzenschutzmittel; Deutsche Forschungsgemeinschaft Ed.; Verlag Chemie, 1985; Methode 6-D.
19. Karlhuber, B. A.; Hörmann, W.D.; Ramsteiner K.A. Anal. Chem. 1975, 47, 2450-2452.

RECEIVED April 16, 1990

Chapter 18

An Enzyme-Linked Immunosorbent Assay (ELISA) for Maduramicin in Poultry Feed

Rosie B. Wong

Agricultural Research Division, American Cyanamid Company, Princeton, NJ 08543–0400

An indirect enzyme-linked immunosorbent assay (ELISA) was developed for maduramicin in poultry feed. The assay utilized polyclonal anti-maduramicin antibody raised in rabbits, maduramicin monoamide with 1,6-hexane diamine-conjugated ovalbumin as the coating antigen, horseradish peroxidase conjugated goat anti-rabbit I_gG and 2,2'azinobis(3-ethylbenzthiazoline) sulfonic acid (ABTS) for quantitation. Standard curves ranging from 0 to 80 ng/mL maduramicin were constructed. The assay did not cross-react with monensin, lasalocid, salinomycin, lincomycin, narasin, chlortetracycline or roxarsone. Broiler feed fortified at 4 to 7 ppm maduramicin were shown to be quantifiable by ELISA at an average recovery of 98.1%. This ELISA method for maduramicin in poultry feed is comparable to the established HPLC-F method.

Maduramicin, the active ingredient in the anticoccidial CYGRO is a polyether ionophore isolated from the fermented culture of <u>Actinomadura</u> <u>yumaensis</u>. The chemical structure is shown in Figure 1. It is a very effective agent for the control of the intestinal parasitic disease, coccidiosis, caused by the protozoan <u>Eimeria</u> spp (1 - 2). The ionophore is administered as a prophylactic in feed. Unlike the other anticoccidials on the market which are generally effective at dose levels of about 60-125 ppm, maduramicin is administered in the range of 5-6 ppm. The present analytical method for maduramicin in poultry feed utilizes a sensitive, high performance liquid chromatography (HPLC) method (3). The method requires the conversion of maduramicin to a fluorescent derivative in order to be quantified by a fluorescence detector on the HPLC system. A radioimmunoassay (RIA) for maduramicin was also developed in the laboratory (4). The assay utilizes tritiated maduramicin as the competing ligand, rabbit polyclonal antibody and a precipitating anti-rabbit I_gG antibody. The assay was useful in tissue residue analyses as well as feed quantitation. However, due to the specialized equipment needed for RIA as well as the requirement for frequent radiosynthesis of the ligand due to radiodegradation, this method proved to be impractical for broad applications.

Enzyme-linked immunosorbent assays (ELISA) have been shown to be viable alternatives to conventional analytical methods. As in all immunoassays the ELISA systems utilize the specificity of the antibodies generated against analytes of interest to selectively seek out these analytes from a heterogeneous matrix for quantitation. ELISA methods use enzymes and color substrates for quantitation. Recent reviews by Mumma et al., (5), Hammock et al., (6) and Lankow et al., (7) presented detailed discussions of the technology, as well as the benefits of immunoassays such as simple, specific, inexpensive, easily automated and highly portable. ELISA methods have been reported for a number of pesticide residues in foods (8 - 12) and animal feed antibiotics (13 - 14). An ELISA method which can detect maduramicin in crude feed extracts would be an alternative method to the existing HPLC-F system. This paper reports our successful development of an indirect ELISA method for maduramicin. The importance of an appropriate coating antigen necessary for the establishment of an ELISA system will be discussed.

Material and Method

Apparatus. Quantitation of color developed in the microwells was achieved by a Vmax Reader (Molecular Devices Corp., Palo Alto, California). Standard curve using semi-log curve fit and unknown calculations were accomplished by using the software program Softmax provided by the manufacturer. The Ultrawasher II (Dynatech Laboratories Alexandria, Virginia) was used to wash the wells.

Reagents. Maduramicin and tritiated-maduramicin were produced by fermentation of Actinomadura yumaensis. Tritium label was obtained by incorporating ^{3}H propionate in the fermentation broth (15). Isolation and purification was carried out at the American Cyanamid Company. Purity was determined by HPLC. Stock solutions at 1 mg/mL or above were prepared in isopropanol, dilutions were prepared in 10% isopropanol. All solutions were stored in glass containers. Bovine serum albumin (BSA), ovalbumin (OA) were from Sigma (St. Louis, Missouri), N-hydroxysulfosuccinimide (sulfo-NHS) and ethyl diamino carbodiimide (EDC) were from Pierce (Rockford, Illinois), 1-cyclohexyl-3(2-morpholinoethyl) carbodiimide metho-p-toluene sulfonate, ethyl chloroformate, diaminohexane were from Aldrich (Milwaukee, Wisconsin), horseradish peroxidase-labeled goat anti-rabbit I$_g$G, enzyme substrate 2,2'azinobis (3-ethyl benzthiazoline sulfonic acid) (ABTS), hydrogen peroxide and protein assay dye solution were from Bio-Rad (Richmond, California). All other reagents were analytical grade.

Synthesis of Maduramicin-Protein Conjugates. The mixed anhydride method (16) with some modifications (Tsou, H. personal communication) was used to prepare the BSA or OA conjugates to maduramicin (M-BSA, M-OA). A small amount of tritiated maduramicin was added to 100 mg of non-radioactive maduramicin resulting in a specific activity of 23 μCi/gm. Fifty microliters of triethylamine was added to the dried maduramicin followed by the addition of ethyl chloroformate. The mixture was allowed to react for 30 minutes at 4°C with stirring. The conjugating protein BSA or OA was dissolved in water followed by the addition of dimethylformamide to a final concentration of 50%. The final concentration of the protein was about 2 mg/mL. The maduramicin anhydride was added to the cold protein solution dropwise with stirring. The molar ratio of maduramicin to protein was 40:1. The reaction was carried out at 4°C for 3 1/2 hours. The product was dialyzed against 0.01 M sodium phosphate buffer pH 7.4 at 4°C until the radioactivity of the dialysate was below 50 dpm/mL. Estimation of the extent of conjugation was accomplished by determining the protein concentration by dye binding (Bio Rad manufacturer's procedure) and the radioactivity using 23 μCi/gm as the specific activity of maduramicin.

Anit-Maduramicin Antibody Production. New Zealand white rabbits were hyperimmunized with maduramicin-BSA in Freund's complete adjuvant and boosted biweekly with Freund's incomplete adjuvant four weeks after the initial immunization. Weekly bleeds were performed. Serum titre was monitored by indirect enzyme immunoassay described below. The antibody was produced through a contract with Hazleton Laboratory (Vienna, Virginia).

Synthesis of Maduramicin Monoamide with 1,6-Hexanediamine (M-C$_6$NH$_2$). One gram of tritium-labeled maduramicin (specific activity 23 μCi/gm) was dissolved in 10 mL of toluene with 100 μL of triethylamine. Ethyl chloroformate (100 μL) was diluted in 15 mL of toluene and was added dropwise to the cold maduramicin solution and allowed to stir at 4°C for 30 minutes. A 10 fold molar excess of 1,6-hexanediamine (1.2 gm) was dissolved in 10 mL of toluene and added. After 1 1/2 hours of reaction at 4°C, ice cold water (4°C) was added to remove salts and clarify the toluene solution. The reaction product was monitored by silica gel thin-layer chromatography and radioactivity. Maduramicin monoamide with 1,6-hexanediamine (M-C$_6$NH$_2$) was purified by silica gel column chromatography using a gradient of methanol in ethyl acetate with or without triethylamine as the mobile phase. The product was stored in toluene at 4°C.

Synthesis of M-C$_6$NH$_2$-Ovalbumin Conjugate (M-C$_6$-OA). Conjugation reaction using N-hydroxysulfosuccinimide (sulfo-NHS) and ethyl diamino carbodiimide (EDC) was employed (17). M-C$_6$NH$_2$ was dissolved in 1 mL of dimethylformamide and added to 30 mL of ovalbumin solution at 40:1 molar ratio of ligand to protein. Sulfo-NHS and EDC were separately dissolved in 5 mL of 0.05 M phosphate buffer pH 7.6 and added to the protein ligand mixture in sequence. The final concentrations of sulfo-NHS and EDC were 5 mM and 50 mM respectively. The reaction solution was stirred at room temperature for 18 hours. Excess reagent was removed by dialysis. Conjugation number was determined by radioactivity (specific activity 23 μCi/gm) and protein concentration.

Indirect Enzyme Immunoassay (ELISA). The indirect ELISA of Engvall et al., (18) was employed with some modification. Coating antigens (M-OA or M-C$_6$-OA) were diluted in 0.01 M sodium phosphate pH 7.6 with 0.14 M sodium chloride and .05% thimerosal (PBS 7.6). Polystyrene 96 well microtiter plates were coated with 100 μL of antigen for 18 to 20 hours at 4°C or 2 hours at room temperature. The coated wells were washed four times with PBS-T (PBS 7.6 containing 0.1% Tween 20) using the Ultrawasher II. The wells were further blocked with 200 μL of 2% ovalbumin in PBS 7.6 or 5% non-fat dried milk (Carnation) in PBS 7.6 for 30 minutes at room temperature. After washing four times with PBS-T, the dried plates were stored sealed at 4°C.

When performing sample analysis or standard curves, 50 μL of antibody and 50 μL of standard maduramicin or sample solution were added to the coated microwells. Incubation was carried out at 4°C for 1 hour. After washing with PBS-T, 100 μL of an appropriate dilution of horseradish peroxidase-conjugated goat anti-rabbit IgG was added and incubated for 1 hour at 4°C. After removing the unbound enzyme by washing, the bound enzyme was quantified by adding 100 μL of freshly prepared substrate ABTS. The color was followed kinetically at 405 nm for 2 minutes or by determining the absorbance at 5 minutes using the Vmax reader. For the determination of serum titre, 100 μL of antibody dilutions in PBS 7.6 were added to the coated wells without the standard maduramicin solution. Incubation and quantitation steps were the same as described above.

Extraction and Analysis of Maduramicin from Poultry Feed. Feed extraction was accomplished by adding 250 mL of acetonitrile to 50 gm of medicated feed. The mixture was shaken for 30 minutes on a reciprocal shaker. A clear aliquot of the sedimented extract was removed for ELISA. At least a fifty fold dilution of the acetonitrile solution was assayed directly without further processing. Standard curves using the corresponding dilution of an acetonitrile extraction of a control feed sample was used to quantitate medicated feed samples. The HPLC-F analysis of the same samples were carried out according to the method described previously (3).

Specificity Study. Compounds used for the specificity study were monensin, lasalocid, narasin, salinomycin, lincomycin, chlortetracycline and roxarsone. The compounds were first dissolved in acetone or acetone:water (8:2 v/v) at a concentration of 1 mg/mL. Further dilutions were prepared in PBS 7.6. Concentrations ranging from 0 to 10 µg/mL were tested. The assays were performed by using 50 µL of the varying concentrations of the compounds in place of the maduramicin in the standard curve procedure as described above. The concentration of the competing compounds required to exhibit 50% displacement of the antibody binding to the solid support was calculated.

Results and Discussions

The addition of radiolabeled ligand in the protein conjugation reactions facilitated the estimation of the extent of conjugation. The BSA conjugate was calculated to contain 7.2 moles/mole of protein; 14 moles per mole of OA was found using maduramicin directly linked through the mixed anhydride method while M-C_6NH_2 conjugation yielded 9.9 moles/mole OA using the sulfo-NHS, EDC reagents.

   The reaction between diaminohexane and carboxyl-activated maduramicin resulted in the production of both mono- and di-substituted hexane molecules. As depicted in the silica gel column chromatogram (Figure 2), both maduramicin and the di-substituted hexane compounds eluted in a solvent of low percentage methanol. However, the mono-substituted M-C_6NH_2 did not emerge until a 2% triethylamine in 80% methanol solvent was used. The yield of this compound was 32.9%.

   Based on the amount of maduramicin or C_6-maduramicin loaded per mole of ovalbumin, equal quantity of maduramicin in M-OA or M-C_6-OA were used as coating antigens. The antibody titers are presented in Figure 3. It is evident that the antibody titre is greater using M-OA as the coating antigen. This is perhaps due to the presence of antibody to the linker between the ligand and protein which is common between the immunogen M-BSA and the coating antigen M-OA. When the two coating antigens were used to construct standard curves, the results are shown in Figure 4. The M-OA standard curve is shallow and the displacement of antibody binding to the solid phase antigen (i.e., coating antigen M-OA) cannot be accomplished by maduramicin alone. On the other hand, the standard curve using M-C_6-OA as coating antigen has a linear response to maduramicin concentration. This confirms that the antibody to the linker between the ligand and the protein for M-BSA is also binding to the M-OA coated on the wells. This binding can not be displaced by maduramicin. The way to improve the maduramicin response was to change the linker between the ligand and the coating protein. It was found that the use of a six carbon bridge on the ligand was sufficient to generate a linear dose response to maduramicin. Thus using M-C_6-OA as the coating antigen, the standard curve was linear between 3 to 80 ng/mL of maduramicin. As shown in Table I, 55 ng/mL of free

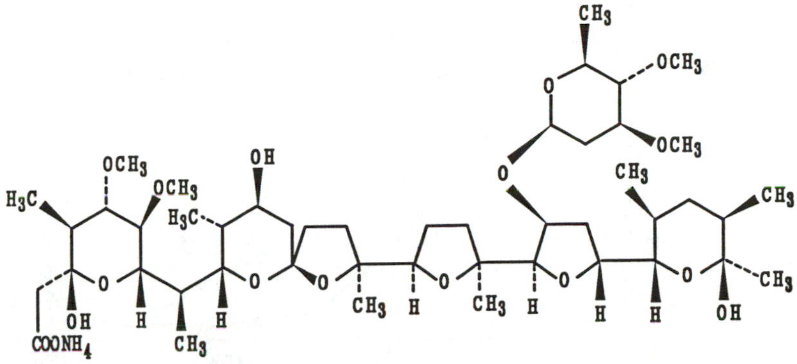

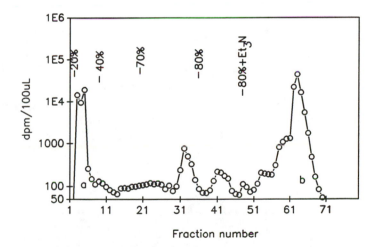

Figure 1. Molecular structure of maduramicin.

Figure 2. Silica gel column purification of maduramicin and diamino hexane reaction product. A 2.5 X 40 cm column of Woelm activity grade silica gel was equilibrated with ethyl acetate. The mobile phase was a step gradient of methanol in ethyl acetate. Arrows indicate the concentration of the methanol used. The last peak was eluted with 80% methanol containing 2% triethylamine. Peak a: unreacted maduramicin and di-substituted hexane diamine. Peak b: mono-substituted hexane diamine.

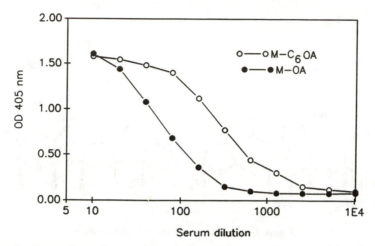

Figure 3. Comparison of serum titre using equal amounts of maduramicin. Coating antigens: maduramicin-ovalbumin (M-OA) and maduramicin monoamide hexane-ovalbumin (M-C$_6$-OA). The enzyme activity is represented as OD 405 nm.

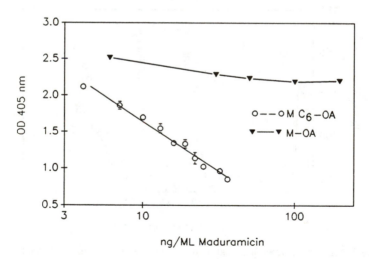

Figure 4. Comparison of Maduramicin standard curves using different coating antigens. Coating antigens: maduramicin-ovalbumin (M-OA) and maduramicin monoamide hexane-ovalbumin (M-C$_6$-OA). On the Y-axis the enzyme activity is represented as OD 405 nm. The X-axis represents the log of maduramicin concentration.

Table I: Inhibition of Maduramicin ELISA (I_{50}) by Common Feed Additives

maduramicin   55 ng / ML

narasin $> 1.0 \times 10^4$ ng /ML

chlortetracycline $> 1.0 \times 10^4$ ng /ML

salinomycin $> 1.0 \times 10^4$ ng /ML

lasalocid $> 1.0 \times 10^4$ ng /ML

monensin $> 1.0 \times 10^4$ ng /ML

lincomycin $> 1.0 \times 10^4$ ng /ML

roxarsone $> 1.0 \times 10^4$ ng /ML

maduramicin was sufficient to displace 50% of the antibody that will bind to the solid supported maduramicin. On the other hand, ionophores such as monensin, lasalocid and feed additives such as roxarsne, chlortetracycline, lincomycin, narasin and salinomycin were not able to displace any antibody at 10 μg/mL, a concentration 200 fold greater than that demonstrated by maduramicin. This indicates that this ELISA system is very specific for the polyether ionophore maduramicin.

The dose response curves performed in PBS 7.6 buffer and in a non-medicated poultry feed extract solution are shown in Figure 5. The two curves are significantly different in optical density responses but the slopes appear to be similar. A preliminary analysis of broiler feed extract by ELISA indicated that a non-medicated feed extact standard curve must be used for computing the values of unknown feed extracts. The same feed extract samples were analyzed by HPLC-F. The comparison is shown in Table II. The data indicates that the ELISA method of analysis for maduramicin in broiler feed correlates well with the HPLC-F method with a Pearson correlation coefficient of 0.973.

Table II: Comparison of ELISA and HPLC-F Analysis of Broiler Feed Extract

Levels of Fortification	ppm by ELISA	Recovery	ppm by HPLC-F	Recovery
4 ppm	3.1, 3.9	87.5%	4.6, 4.6	115.0%
5 ppm	4.9, 5.1	100.0%	5.3, 5.5	108.0%
6 ppm	6.2, 5.9	100.8%	6.6, 6.8	111.6%
7 ppm	7.2, 7.4	104.3%	7.4, 7.6	107.1%
Average Recovery		98.1%		110.2%

Due to the parallel nature of the buffer and the non-medicated feed extract standard curves, a method of quantification in the absence of a non-medicated feed standard curve is described. This method will require the analysis of each sample twice - once with the addition of a known amount of maduramicin added to the sample and once without the addition. Calculating the values based on a buffer standard curve, the true value of the sample can be generated from the following equation:

$$X(f) = a$$
$$X(f) + k(f) = b$$

Where X is the true value of unknown sample
   a - is the value of feed extract alone based on the buffer standard curve.
   b - is the value of feed extract plus a known quantity of maduramicin based on the
         buffer standard curve.
   f - is the difference factor between the buffer standard curve and the non-medicated
         feed extract standard curve.
   k - is the known quantity of maduramicin added 00to the sample.
from the above equations solving for X

$$X = ak/b-a$$

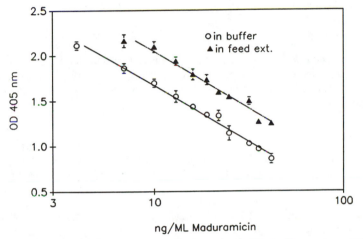

Figure 5. Comparison of maduramicin standard curves. The buffer used was PBS 7.6, the feed extract was a 1:50 dilution of an acetonitrile extract of non-medicated feed extracted as described in the methods section and diluted in PBS 7.6.

Summary

We have shown an ELISA method for maduramicin using a polyclonal antibody against a mixed anhydride conjugation of maduramicin with bovine serum albumin as immunogen and an ovalbumin conjugate of a modified maduramicin molecule ($M-C_6NH_2$) as the coating antigen. The maduramicin dose response is linear between 3 to 80 ng/mL. The specificity of the antibody is good with respect to common feed additives. The feed assay correlates well with the established HPLC-F method. Analysis of the feed extract in the laboratory is based on the non-medicated feed extract standard curve. In the event that non-medicated feed is not readily available, a buffer standard curves may be used for quantitation using the fortification method discussed above. Validation of the method should be performed before it can be used for field sample analysis.

Acknowledgment

I thank Dr. M. Bullock, H. Tsou and Z. Ahmed for the tritiated maduramicin as well as for guidance in the derivatization methodology, A. Kleiner and Dr. M. Blank for providing the antisera, A. Markantonatos for the HPLC-F analysis, and Y. Roman for her skillful technical assistance.

Literature Cited

1. Kantor, S.; Schenkel, R. H. Poultry Science **1984**, 63, 1497 - 1505.
2. Kantor, S.; Schenkel, R. H.; Kennett, R. L. Jr. Poultry Science **1984**, 63, 1506 - 1511.
3. Markantonatos, A. J. Liq. Chrom. **1988**, 11, 877 - 890.
4. Kleiner, A. Report M-1937, **1983**, American Cyanamid Agricultural Research Division, Princeton, New Jersey.
5. Mumma, R. O.; Brady, J. F. Sixth IUPAC International Congress of Pesticide Chemistry Pesticide Science and Biotech **1986**, 341 - 348.
6. Hammock, B. D.; Gee, S. J.; Cheung, P. Y. K.; Miyamoto, T.; Goodrow, M. H.; Van Emon, J.; Seiber, J. N. Sixth IUPAC International Congress of Pesticide Chemistry Pesticide Science and Biotech, **1986**, 309 - 316.
7. Lankow R. K.; Grothaus, G. D.; Miller S. A. ACS Symposium Series 334, Biotechnology in Agricultural Chemistry, **1987**, 228 - 252.
8. Wie, S. I.; Hammock, B. D. J. Agri. Food Chem. **1982**, 30, 949 - 957.
9. Wie, S. I.; Hammock, B. D. J. Agri. Food Chem. **1984**, 32, 1294 - 1301.
10. Kelley, M. M.; Zahnow, E. W.; Peterson, W. C.; Toy, S. T. J. Agri. Food Chem. **1985**, 33, 962 - 965.
11. Niewola, Z.; Benner, J. P.; Swaine, H. Analyst **1986**, 111, 399 - 403.
12. Newsome, W. H. Sixth IUPAC International Pesticide Congress, Pesticide Science and Biotech **1986**, 349 - 352.
13. Heitzman, R. T.; Carter, A. P.; Cottingham, J. D. Br. Vet. J. **1986**, 142, 516 - 523.
14. Mount, M. E.; Failla, D. S. J. Assoc. Off. Anal. Chem. **1987**, 70, 201 - 205.
15. Tsou, H.; Rajan, S.; Fiala, R.; Mowery, P. C.; Bullock M.; Borders, D. B.; James, J. C.; Martin, J. H.; Morton, G. O. The J. of Antibiotics **1984**, 37, 1651 - 1663.
16. Erlanger, B. F.; Borek, F.; Beiser, S. M.; Liebman, S. J.B.C. **1959**, 234 1090 - 1094.
17. Staros, J. V.; Wright, R. W.; Swingle, D. M. Anal. Biochem. **1986**, 156, 220 - 222.
18. Engvall, E; Perlmann, P.; J. of Immuno. **1972**, 109 129 - 135.

RECEIVED April 16, 1990

Author Index

Berkowitz, David B., 15
Carlson, Robert E., 95
Carpino, L. A., 140
Clarkson, Carolyn E., 95
Dargar, Ratna V., 170
Deschamps, Raymond J. A., 66
Eck, D. L., 79
Egger, Marie L., 95
Feng, Paul C. C., 180
Föry, Werner, 199
Frevert, Jürgen, 193
Gee, Shirley J., 112
Goodrow, Marvin H., 112
Hall, J. Christopher, 180
Hammock, Bruce D., 95
Harrison, Robert O., 112
Horton, Susan R., 180
Jacobs, Jeffrey W., 95
Jung, Freia, 112
Karu, Alexander E., 95
Krogsrud, Stephen, 21
Kurth, M. J., 79
Lang, Kenneth T., 21
Li, Qing Xiao, 112,156
Logusch, Eugene W., 180

Lucas, Anne D., 112
Macmillan, C., 79
Mei, J. V., 140
Mumma, Ralph O., ix
O'Rangers, John J., 27
Page, Samuel W., 38
Pohland, Albert E., 38
Ramsteiner, Klaus, 199
Reck, Bernhard, 193
Schlaeppi, Jean-Marc, 199
Schmidt, Douglas J., 95
Seiber, James N., 156
Sharp, C. Ray, 180
Stoddard, Peter J., 51
Sundaram, K. M. S., 112
Swanson, Todd A., 95
Székács, András, 112
Trucksess, Mary W., 38
Tymonko, John M., 170
Van Der Werf, Paul, 170
Van Emon, Jeanette M.,58,95,156
Van Vunakis, Helen, 2
Wong, Rosie B.,211
Wratten, Stephen J., 180
Yin, C.-M., 140

Affiliation Index

American Cyanamid Company, 211
Batelle-Institut e.V., 193
Brandeis University, 2
California Department of Food and
    Agriculture, 51
Ciba–Geigy, Ltd., 199
ECOCHEM Research, Inc., 95
FMC Corporation, 170
Monsanto Agricultural Company, 180
Pennsylvania State University, ix
Sonoma State University, 79

U.S. Army Toxic and Hazardous Materials
    Agency, 21
U.S. Department of Agriculture, 15
U.S. Environmental Protection Agency,
    58,95,156
U.S. Food and Drug Administration,
    27,38
University of California–Berkeley, 95
University of California–Davis,
    79,112,156
University of Guelph, Ontario, 66

Subject Index

A

Abamectin, use in insecticides, 95
Activated esters of methoprene, synthesis, 146-150
Aflatest immunoaffinity column, effectiveness, 45,47t,48
Aflatoxins
  commercially available immunoassay test kits, 41-49
  general structure, 38,39f
Agrochemicals, solubility problems for immunoassays, 4
Alachlor
  ELISA, use for detection, 180-191
  GC-MS analytical procedure, 181
  structure, 180,185f
Amitrole, hapten attachment strategies, 130,132f
Analyte, source, 2
Analytical methods of monitoring environmental pollutants
  immunochemical methods, advantages, 79-80
  importance, 79
  limitations, 79
Analytical procedure, assessment of potential usefulness, 3,5
Animal drug residue immunoassay development
  advantages, 28
  intended use, effect on performance evaluation of method, 29-30
  regulatory situation, 28-30
  reliability, 28
  technical considerations, 30-37
Antibody
  antibody-binding fragments, 10
  characterization description, 31
  definition, 5
  idiotypic antibodies, 10-11
  mechanism of production, 8
  monoclonal antibodies, 10

Antibody-*Continued*
  polyclonal antibodies, 8,10
  production, 5
  production in genetically engineered bacteria, 9
  production techniques, 2
  rabbit antibodies, 8,9f
  single-domain antibodies, 10
  specificity, 5,7-8
  stereospecific antibodies, 10
Antibody-based veterinary diagnostic tests, market, 27
Antigen, description, 5,6f,7
Antigen-binding fragments, comparison to antibody molecules, 10
Antiidiotypic antibodies
  applications to biological problems, 8
  production, 11
Atrazine
  detection methods, 199-200
  use as herbicide, 199
Avermectins
  detection with monoclonal antibodies, 96-110
  discovery, 95
  extraction difficulties, 96
  structure, 96,97f
  use as cidal agents, 95

B

Bacillus thuringiensis kurstaki endotoxin, immunoassay development, 133,136f
Bentazon
  advantages and disadvantages of ELISA, 164,166
  approaches to obtain hapten-protein conjugate, 164t
  ELISA optimization, 164,165f
  hapten attachment strategies, 130,132f

Bentazon–*Continued*
relative sensitivity of rabbit antibody,
130,131*f*
use as herbicide, 164
Bioassays, measurement of function and
structural identity, 32
Biochemical methods of analysis, growth
in routine applications, 27

C

California Department of Food and
Agriculture, development and use of
immunochemical assays, 51-56
Clomazone
ELISA analysis, 171-178
structure, 173,174*f*
use as herbicide, 171
Clopyralid, chemical structure, 70,73*f*
Committee for the Evaluation of Immunoassay in Environmental Chemistry
possible roles, 125-126
problems, 126
Competition immunoassay for endosulfan
detection
characteristics, 194*t*
cross-reaction
chlorohydrocarbons, 196,198*t*
endosulfan derivatives, 196*t*
development, 194
limitations, 195
procedure, 195
standard curves, 195,197*f*
Cotinine
antibody production, 9
formation during nicotine metabolism, 5
half-life, 5

D

Developmental criteria for immunoassays
availability
confirmatory data, 63

Developmental criteria for immunoassays
availability–*Continued*
cost estimates, 63
reagents and instruments, 63
cross-reactivity characterization of
specific antibody, 62
environmental matrices, characterization
of effects on assay performance, 62
hapten synthesis and conjugation
procedures, documentation, 62
immunization schedules and fusion
protocols, documentation, 62
optimized method, 62
standard operating procedure, 62
well-defined data quality objectives, 62
Drug residue immunoassays, technical
considerations in development, 27-37

E

Endosulfan
competition and inhibition immunoassays,
use for detection, 193-198
immunoassay development requirements, 193
Environmental immunoassays, evaluation
studies, 58-63
Enzyme-linked immunosorbent assay(s)
(ELISA)
detection of nitroaromatic compounds,
80-93
meat inspection, 16
Enzyme-linked immunosorbent assay for
alachlor detection
alachlor protein conjugates, synthesis,
181-182
antibody generation, 182-183
checkerboard assay procedure, 182
cross-reactivity of antibodies,
184,186-189
chloroacetanilide, 184,186,187*f*
thioether analogues, 186,188*f*
cross-reactivity study procedure, 183

Enzyme-linked immunosorbent assay for alachlor detection–*Continued*
environmental water samples, analysis, 189,190*f*,191
experimental instrumentation and materials, 181
immunoassay procedure, 182-183
inhibition analysis in deionized water, 184,185*f*
maximization of epitope density, 184,185*f*
Enzyme-linked immunosorbent assay for clomazone detection
antiserum preparation, 171
clomazone residual soil levels by ELISA vs. wheat injury from greenhouse bioassay, 175,176*f*
comparison to GLC technique, 175,176*f*
distribution of corn hybrid sensitivity toward clomazone, 175,177*f*,178
ELISA procedure, 172
greenhouse bioassay procedure, 172
protein conjugate preparation, 171
reproducibility, 173
soil extraction procedure, 172
soil recovery, 173,174*f*
specificity, 173
Enzyme-linked immunosorbent assay for maduramicin detection
antibody titers, 214,216*f*
antimaduramicin antibody production, 213
comparison of ELISA and HPLC analy - sis of broiler feed extract, 218*t*
dose-response curves in buffer, 218,219*f*
experimental apparatus and reagents, 212
extraction and analysis of maduramicin from poultry feed, 214
indirect ELISA procedure, 213-214
inhibition by feed additives, 214,217*t*,218
maduramicin monoamide, synthesis with 1,6-hexanediamine, 213

Enzyme-linked immunosorbent assay for maduramicin detection–*Continued*
maduramicin monoamide-1,6-hexanediamine-ovalbumin conjugate synthesis, 213
maduramicin-protein conjugate synthesis, 212-213
quantification method without nonmedicated feed standard curve, 218
silica gel column of maduramicin and diaminohexane reaction product, 214,215*f*
specificity study procedure, 214
standard curve vs. coating antigen, 214,216*f*
Enzyme-linked immunosorbent assay for methoprene detection
assay development, 150-151,152*f*
cross-reactivity of methoprene antiserum with methoprene derivatives and juvenile hormones, 150-151,152*f*
immunogen synthesis, 143-150
methoprene extraction from tobacco, 151,153*f*,154
methoprene immunogen, design, 142-143,144*f*
percent inhibition of antiserum by methoprene, 150,152*f*

F

Fluroxypyr, chemical structure, 70,73*f*
Food Safety and Inspection Service
detection of pathogens and chemicals in food supply, 16
function, 15
procedure for meat inspection, 15
test review process and criteria, 16-20
Food safety applications, immunoassays, 38-49
Food supply, detection of pathogens and chemicals, 16

G

Glyphosate, immunoassay development,
130,133,134f

H

Hapten, 4
preparation of derivatives for conjugation to macromolecules, 5,6f
specific antibody source, 5,7
Hapten tracer method, description, 54-55
Hydroxyatrazine, use of monoclonal antibody based enzyme immunoassay for detection, 200-209
N-Hydroxysuccinimide ester of methoprene, synthesis, 146,147f

I

Idiotype-antiidiotype interaction, applications to biological problems, 8
Idiotypic antibodies, production, 10-11
Immunoassay
advantages
agricultural applications, 170
meat inspection, 16
pesticide analysis, 180
analyte source, effect, 3,5
antibodies, 7-12
antigens, 5,6f,7
development, 1
sensitivity, 2
specificity, measurement, 35
Immunoassay for pesticide analysis
advantages, 156-157,180
amount of information delivered, 157
bentazon, 164t,165f,166
disadvantages, 157,159
examples, 157,158t
inexplicable pitfalls, 157,159
method development time, 157
molinate, 161,162-163t,164
paraquat, 159,160f,161t

Immunoassay in food safety applications
Aflatest immunoaffinity column,
45,47t,48
aflatoxin(s), general structure, 38,39f
aflatoxin test kits
commercially available, 41,42t
operating characteristics curves, 48,49f
Immuno Dot Screen Cup, 45,46t,47f
meat inspection, 180
Neogen Agri-Screen kit, 41,43-44t,45
screening method, criteria for evaluation, 40
Immunoassay methods for nitroaromatic compounds
active N-hydroxysuccinimide esters, preparation, 82-83
antisera, 82
antisera dilution factors for coating antigens, comparison, 84,86t
checkerboard titration of antisera and coating antigen, 87,91
checkerboard titration procedure, 84,85f,86t
comparative inhibition, calculation, 87,88t,89-90f
competitive ELISA with nitroaromatic standards, procedure, 84,87,89-90f
cross-reactivity for analytes on different antisera-coating antigen systems, 87,88t
detectable compounds, trends, 92-93
ELISA, sensitivity to analytes, 91-92
esters of haptens, structures, 80,81f,82
general enzyme-linked immunosorbent assay procedure, 83-84
hapten-protein antigens, synthesis, 87
immunizing and coating antigen, preparation, 83
nonspecific binding procedure, 84
protein conjugates, synthesis, 82,85f
reagents, 82
reproducibility, 92
sensitivity and precision, maximization, 92
synthesis strategy, 80,81f,82

Immunochemical assay, development and use by California Department of Food and Agriculture, 53-55,54*t*
hapten tracer method, 54-55
history, 51,52*t*,53
pesticide residues in groundwater, 52*t*
problems to overcome, 55-56
targets for future antibody production, 53-54*t*
Immunochemical methods, advantages for use in monitoring environmental pollutants, 79-80
Immunochemical technology in environmental analysis
advantages and limitations, 113
analyst skill, required, 114-115
antibody source, problems, 114
applicability to analytical problems, 115,116*f*,117*t*
assay development time line, 128,131*f*
avoidance of duplication, 124-125
changes in last 10 years, 112-113
choice of format, 118-119
choice of hapten, 130
Committee for the Evaluation of Immunoassay in Environmental Chemistry
possible roles, 125-126
problems, 126
committee to coordinate development, 125
comparison to bioassays, 114
development
biotechnology companies, 120-121
immunoassays for environmental analysis in other fields, 115
in-house, 119-120
universities and government agencies, 120
evolution of problems, 112-115
hapten design, 129-130
immunoassay area, recent work, 126–127
importance of analyte-antibody interaction, 130,133,134-136*f*
jargon, problems, 114
misconceptions, 113–115

Immunochemical technology in environmental analysis—*Continued*
molinate assay development example, 128-129
monoclonal vs. polyclonal antibodies, use, 121-122
properties of compounds affecting ease of assay development, 117*t*
proprietary information, disclosure, 119
qualitative vs. quantitative formats, 118
role of industries and regulatory agencies in advancement, 123–124
standardization, 122–123
water sample assay development, example, 129
Immunochemistry, use for pesticide residue analysis, 67
Immuno Dot Screen Cup
collaborative study results, 45,46*t*,47*f*
description, 45
Immunoglobulin G molecule, schematic structure, 8,9*f*
Inhibition immunoassay for endosulfan detection
characteristics, 194*t*
cross-reaction
chlorohydrocarbons, 196,198*t*
endosulfan derivatives, 196*t*
development, 193-194
procedure, 195
standard curves, 196,197*f*
Insect growth regulators, function, 140
Ivermectin
structure, 96,97*f*
use in insecticides, 95
use in treatment of river blindness, 95-96

J

Juvenile hormone, structure, 140,141

M

Maduramicin
detection methods, 211
ELISA analysis, 212-219

Maduramicin–*Continued*
  function, 211
  structure, 211,215f
Methoprene
  enzyme-linked immunosorbent assay for
    detection, 142-154
  structure, 140,141f
Methoprene immunogen, design,
  142-143,144f
Methoprene immunogen synthesis
  comparison of hapten density by
    different active ester methods, 149,150t
  incorporation of spacer group,
    143,145f,146
  number of methoprene molecules intro-
    duced per protein molecule, 149t
  synthesis of activated esters, 146-150
Molinate
  ELISA
    advantages and disadvantages, 163-164
    inhibition by thiocarbamates, 161,162t
    standard deviation of standard curves,
      161,163t
  recovery from spiked water by ELISA
    and GC, 161,163t
  structure, 161
  use as herbicide, 161
Monoclonal antibody
  advantages over polyclonal antibodies,
    67
  definition, 67
  production from single clone of B cell, 8
  use in production of stereospecific
    antibodies, 8
Monoclonal antibody-based enzyme
    immunoassay for atrazine and
    hydroxyatrazine
  atrazine and hydroxyatrazine conjugates,
    synthesis, 200,201f
  atrazine determination in soil, 208,209f
  competitive ELISA procedure, 200,202
  cross-reactivity of antibodies, 203,204t
  hydroxyatrazine determination in soil,
    206,207f
  immunization and fusion procedure, 200

Monoclonal antibody-based enzyme
    immunoassay for atrazine and
    hydroxyatrazine–*Continued*
  monoclonal antibody characterization,
    202t
  recovery of hydroxytriazines from
    fortified soils, 203,205f,206,207f
  soil sample analytical procedure, 202
Monoclonal antibody detection of
    avermectins
  advantages, 105
  domain effects, 109
  enzyme immunoassay procedure,
    99,100f
  extraction procedure, 105,106f
  hybridoma preparation, 98
  molecular models of avermectins,
    102,104f,105
  problems to overcome, 110
  recovery and analysis of residues, 98-99
    from spiked water, 105,107t
    from strawberry homogenates,
      105,108t
    procedure, 105,106f
  reproducibility, 99,101f,102
  sensitivities, 105,109
  specificities of antibodies, 102,103t
  structural effects, 109
  synthesis of haptens and conjugates, 96
Monoclonal antibody technology program
    of U.S. Army Toxic and Hazardous
    Materials Agency
  advantages, 22
  compounds selected for immunoassay
    development, 22,23f
  developmental approach, 22
  experimental procedure, 22
  hybridoma production, 25t
  structures of immunogens, 22,24f
  tetryl monoclonal antibody
    cross-reactivity, 25t
Monoclonal immunoassay for picloram
    detection
  antipicloram antibody, production,
    69-70,73f

Monoclonal immunoassay for picloram
    detection—*Continued*
  coating conjugates, preparation, 69
  determinations for radioimmunoassays,
    accuracy, 72t,74
  experimental chemicals and materials,
    68
  functions, 77
  immunogens, preparation, 68
  indirect enzyme immunoassay proce-
    dure, 71-72
  intraassay variability for indirect enzyme
    immunoassays, 74t,75
  radioimmunoassay procedure, 72
  radiolabel, preparation, 68
  recovery from fortified human urine
    samples by enzyme immunoassay,
    75,76t
  sample preparation, 70-71
  standard curves
    indirect enzyme immunoassays,
      73f,74
    radioimmunoassay procedure, 72,73f

N

Neogen Agric-Screen kit
  collaborative study results,
    41,43-44t,45
  content, 41
  procedure, 41
Nicotine
  antibody production, 9
  metabolism pathways, 3,4f
Nitroaromatic compounds
  frequency of occurrence, 80,81f
  immunoassay methods for monitoring,
    80-93
  need for extensive monitoring, 80
  structures, 80,81

P

Paraquat
  analytical method, 159
  ELISA, procedure, 159,160f
  ELISA and spectroassay, detection
    limits, 161t
  immunoassay analysis, barriers, 159,161
  structure, 159
  use as herbicide, 159
Pesticide analysis,
  cost and time, 156
  immunoassay(s)
    advantages, 67,156-157
    disadvantages, 157,159
    examples of pesticides, 157,158t
Picloram
  chemical structure, 70,73f
  monitoring requirement, 66-67
  polyclonal and monoclonal immunoas-
    says for detection, 67-77
  role as herbicide, 66
Polyclonal antibody
  definition, 67
  factors affecting choice, 8,10
Polyclonal immunoassay for picloram
    detection
  antibody production, effect of immuno-
    gen, 77
  antipicloram antibody, production, 69
  coating conjugates, preparation, 69
  determinations for radioimmunoassays,
    accuracy, 72t,74
  experimental chemicals and materials, 68
  functions, 77
  immunogens, preparation, 68
  indirect enzyme immunoassay proce-
    dure, 71-72
  intraassay variability for indirect enzyme
    immunoassays, 74t,75
  radioimmunoassay procedure, 72
  radiolabel, preparation, 68
  recovery from fortified water, soil, and
    plant samples by enzyme immunoas-
    say, 75t,76

Polyclonal immunoassay for picloram
detection–*Continued*
sample preparation, 70-71
standard curves
indirect enzyme immunoassays,
73*f*,74
radioimmunoassay procedure, 72,73*f*

R

Radioimmunoassay
development, 27
role in meat inspection, 16
River blindness, use of ivermectin in
treatment, 95-96

S

Screening method, definition, 40
Screening test development, technical
considerations, 30-37
Single-domain antibodies, production, 10
Stereospecific antibodies, production, 8
Synthetic organic chemicals, monitoring
techniques, 79

T

Technical considerations in screening test
development
assay sensitivity with replication and
statistics, determination, 33-34
intended use of method, 30-31
interferences, 35-36
scientific principles and critical reagent
or instruments, description, 31-32
stability data, 33
synthesis and characterization of
critical reagents, method, 32-33
test samples, 35
test system logistics, 33

Technical considerations in screening test
development–*Continued*
validation studies, 36-37
variability associated with each standard
point on analytical curve, 34-35
Testing systems, 20
Test Review System of Food Safety and
Inspection Service
decision(s)
to accept laboratory characterization
of test, 17
to accept results of collaborative study,
19
to approve test, 19-20
to review step, 16-17
evaluation of yes/no tests, 17,19
quantitative vs. qualitative assay
interpretation, 17,18*f*
Tobacco, extraction of methoprene using
competition ELISA, 151,153*f*,154
Triazines, immunoassay development,
133,135*f*
Triclopyr, chemical structure, 70,73*f*

U

U.S. Army Toxic and Hazardous Materials Agency
cleaning up environmental problems, 21
function, 21
monoclonal antibody technology
program, 21-25
U.S. Environmental Protection Agency
evaluations of immunoassay methods
developmental criteria, 62-63
evaluation procedure, 59
guidelines, 59,60*f*,61
importance, 58-59
sample selection, 59
steps, 59,60*f*

W

Water-soluble activated ester of
methoprene, synthesis, 146,148*f*